A Rational Species

Essays on the Environment

Edited by Melissa Gish

with an introduction by Rico M. Gazal, PhD

A Rational Species: Essays on the Environment

Direct inquiries/permissions to

Department of Language and Literature
Glenville State College
200 High Street
Glenville, West Virginia 26351

Library of Congress Control Number: 2020950084
ISBN: 978-1-953033-00-0

Published by Candy Bat Books
United States of America

"Life is life's greatest gift. Guard the life of another creature as you would your own because it is your own. On life's scale of values, the smallest is no less precious to the creature who owns it than the largest."

— Lloyd Biggle Jr., writer and oral historian

CONTENTS

Introduction	Rico M. Gazal	i
Considering Wastewater Pollutants	Taylor Adkison	1
Deep Trouble: Plastic Pollution in Ocean Trenches	Kyra Bauer	5
Species at Risk: Australia's Greater Bilby	Kylie Davis	11
The Great Barrier Reef: Extinction of an Ecosystem?	Jonathan Dodson	15
Leatherback Sea Turtles: Gentle Ocean Giants	Abigail Dye	23
African Elephants and the Ivory Trade	Bradley F. Hall	31
Getting to Know the Arctic Fox	Jessica R. Jenkins	37
The Growing Threat of Hurricanes	Shayne Miller	41
Species at Risk: Grauer's Gorilla	Oluwatobi Oladapo	53
Appalachian Red Spruce Forests	Adam L. Osborne	67
Species at Risk: Virginia Big-eared Bat	Jazmon Pennington	77
Effects of Light Pollution on Wildlife	Lori K. Ray	85
Fabulous Frogs of the World	Brogan Richards	94
What's Next for the Polar Bears?	Heather Sears	99
The California Condor: A Remarkable Recovery	Paula Snyder	105
Sea Turtles: Navigating a Changing World	Trey Waycaster	113
Image Credits		124
Glossary		125

INTRODUCTION

The world around us is faced with many environmental issues and problems that we hear and read about every day. Unless our daily lives are directly affected by environmental problems, we tend to forget that these problems exist because we neglect to pay attention and do our part to address them.

Every ecosystem is unique. Some ecosystems with rare features and characteristics are set aside for protection. They have the most productive soil and water that support a diversity of plants and animals. A majority of these ecosystems are protected by laws, but even with the most restrictive rules and regulations, they are still vulnerable to any kind of changes around them. Environmental laws are meant to protect the integrity of our ecosystems, and if properly implemented, they can make a significant difference in our lives and the lives of plant and animal communities.

Environmental resiliency is the capacity of the environment to recover from disturbances and adapt to changing environmental conditions. Changes in the environment are inevitable. Each ecosystem because of its own unique conditions responds to changes differently. The stability of the ecosystem dictates how fast or slow its recovery after a disturbance. Stable communities tend to recover quickly; however, there are certain ecosystems that are fragile and will recover slowly to changes. Certain ecosystems may not necessarily go back to their original state after a disturbance or will take an unusually longer period to even be anywhere close to where they were before the disturbance. There are also circumstances when an ecosystem that is completely altered has no choice but to adapt to environmental changes. The strategies of ecosystems to recover or adapt to changes may differ, and communities of plants and animals that depend on them may not make it.

The interaction of organisms with an environment is a dynamic process. Disappearance of life due to continuous change in the environment occurs at an alarming rate. Before these communities even respond to one change, another change may come, making the recovery even more difficult. Some species that fail to adapt and eventually leave the environment create further problems as their absence also changes the dynamic interaction in the environment. Some species, as a form of their adaptation to the environment, modify their morphological and physiological characteristics that enable them to adjust to the new environment, but for some, their only escape is to migrate to a different environment. Species migration in response to climate change is a common phenomenon for both plants and animals. Plant species that are known to be confined only in the southern range are moving northwards where warming is taking place. The same is true for more northern species that go farther north in search of a cooler climate.

Disturbances in nature could be light, mimicking changes that are inherent to the environment, or they could be severe, causing long-lasting consequences that can impact recovery. Take for example how invasive plants can colonize a disturbed forest area and make it their own. Continuous encroachment of non-native invasive plants accompanied by their modification of the site conditions pose a great threat to the existence of native plants and to the overall productivity and stability of the forest ecosystem. Understanding the invasive characters of these different invasive plants and the environment where they exist is necessary to determine their impacts on the ecology and functions of our forest ecosystem. Aside from crucial environmental factors for growth and development of exotics, their success is most likely due to their strong competitive abilities (i.e. allelopathy), faster growth rate, and abundant seed crops. It has also been suggested that some of the invasive plants (e.g. *Ailanthus altissima*) once established can modify soil microbial communities and biogeochemical cycling in ways that can feedback to benefit themselves. Increased forest disturbances due to land use change (e.g. urbanization, roads, fire) accompanied by increased availability of sunlight, soil moisture, and nutrients could lead to more opportunities for invasive plants to invade, interact with native species, and potentially have a negative impact on the native plant communities.

Environmental sustainability is a concept that allows us to responsibly manage natural resources to avoid their degradation or depletion and maintain their long-term health. As stewards of these resources, humans are expected to use these resources wisely and ensure that future generations are also able to enjoy these resources. The majority, if not all, of human activities have negative impacts on the environment and are considered to be the major cause of global climate change. Because of these human activities, we have altered our environment in ways that put us all in danger, including those who are yet to be born—the future citizens of the Earth. If nothing changes, we will be depriving them of the joy of nature and its abundant resources. Ecosystems are composed of biotic and abiotic factors that are interconnected and interdependent in performing their functions. This complex interaction in nature drives the flow of energy and nutrients that sustain life. Disrupting this interaction can cause significant local, regional and global impacts. Humans are important components of the ecosystem and have the main responsibility to protect, conserve, and preserve these resources. We must all act together to support environment-friendly projects and actions that aim to enhance our environment's long-term health and sustainability.

Dr. Rico M. Gazal,
Department of Land Resources
Glenville State College

CONSIDERING WASTEWATER POLLUTANTS

Taylor Adkison

"Sewage works that serve big cities run into trouble when the cities grow up around them."

— Rose George, *Ninety Percent of Everything*

Wastewater—water produced by domestic households, industrial, and agricultural practices—can cause pollution of many lakes and rivers. "Sewage" is the term that is used for wastewater that often contains **feces**, urine, and laundry waste.[1] Wastewater that enters into our ponds, streams, lakes, rivers, and oceans affect wildlife, humans, and the environment as a whole. Fish are affected by direct **contamination**. Humans can suffer from contamination by eating the fish, swimming, or directly from wastewater getting into main waterways. In addition, a number of diseases and **pathogens** affect the environment through soil. It is important to understand how wastewater pollution affects us and our world and how we can address problems with wastewater pollution.

Untreated wastewater is a serious problem for the planet. According to the Natural Resources Defense Council (NRDC), "some 80 percent of the world's wastewater is dumped—largely untreated—back into the environment, polluting rivers, lakes, and oceans."[1] Disposal could be purposeful or it could be accidental. Wastewater pollution can happen when a hurricane or a flood occurs. Water being cleaned in wastewater treatment facilities may overflow, forcing wastewater that is not yet treated to mix with clean water that comes from the rain. Once the water is contaminated, it infects all the other water with which it comes into contact. Wastewater may contain heavy metal pollutants that are harmful to humans and wildlife. These **toxic** pollutants may include lead and mercury.

Figure 1 Wastewater treatment plant

When humans or animals ingest or inhale heavy metal pollutants, they can suffer kidney damage as well as gastrointestinal disorders, tremors, inflammation, and blood disorders. Diseases of the nerves, depression, and pneumonia can develop from breathing in toxic **vapor**.[2] Wastewater pollution affects fish by leading to reduced growth and development, cancer, organ damage, nervous system damage, and, in extreme cases, death.

Wastewater management plants that specialize in the treatment of wastewater and laws that require industry and urban areas to treat wastewater are vital to the health of the planet and its inhabitants. Unfortunately, some plants cut corners to trim costs, which leads to failures in proper water treatment. Domestic wastewater management guidelines are not laws; they are simply "recommended" and haven't been adequately revised to reflect population growth or land development since 2004. When wastewater isn't properly cleaned, dangerous **bacteria**, molds, and a wide variety of other microorganisms can take over.

Another problem caused by wastewater pollutants is eutrophication of freshwater **habitats**. Eutrophication is "excessive richness of **nutrients** in a lake or other body of water, frequently due to runoff from the land, which causes a dense growth of plant life and death of animal life from lack of oxygen."[2] **Commercial** agriculture is a major cause of eutrophication. **Organic** and **chemical** fertilizers, feedlot waste, and other agricultural waste products as well as sediment may soak into the soil and make their way to water or run directly into water systems such as ditches and wetlands.

Figure 2 Eutrophication can destroy a fresh water habitat.

FLORIDA: A CASE STUDY

Consider a food chain in Florida's wetlands. In the safety of free-floating bladderwort, mosquitos, water striders, and other insects hatch. These insects feed southern leopard frogs and eastern mud turtles. The frogs and turtles are snatched up by American alligators. Great blue herons grab alligator hatchlings that stray too far from their group and pluck fish such as grass carp from among the drifting bladderwort. A variety of plants aerate the wetland, providing oxygen and offer food and habitat for many **species**. Herons and egrets depend on the fish that swim among the bladderwort and other important wetland plants. Now dump feedlot or commercial fertilizer runoff into this wetland. Sediment is forced into the wetland, lowering the water level. Invasive algae grow, smothering the bladderwort and other plants. The water becomes **stagnant**. Excessive sediment results in the evaporation of the water, and the wetland soon becomes less wet. Insects, frogs, and turtles cannot reproduce. Fish die due to lack of oxygen in the water. Alligators and birds are forced to find new homes or risk starvation.

One thing water treatment plants can do is improve their systems to provide better efficiency and include techniques that take out more nitrogen, (a chemical that causes eutrophication) than current models. What can we do to help stop wastewater pollution? We should dispose of chemicals and automotive fluids and fuels properly, which means not pouring them down the drain or dumping them off the edge of the curb where they will run into storm drains. We should not put fats, oils, or grease down any drain. We shouldn't run automatic dishwashers or clothes washers when it is raining; this will help prevent water treatment facilities from overflowing due to all the water coming into the systems at one time. People using a septic system should get their system inspected regularly and have the tank pumped out as needed to avoid damage and leaks in the system. We should never dispose of chemicals or other hazardous waste in toilets or sinks.

The fresh water on our planet is limited. It is vital that we protect it for now and for the future. When it comes to wastewater pollution, we can be doing a lot more than we are currently doing, from supporting laws that limit or eliminate agriculture runoff and improving wastewater treatment systems to practicing mindful use of our water systems in our homes. It's up to us to protect our wetlands and waterways from the hazards of modern life.

Figure 3 It's up to everyone to protect our fresh water.

1 Denchak, Melissa. "Water Pollution: Everything You Need to Know." *Natural Resources Defense Council (NRDC)*, 14 May 2018, www.nrdc.org/stories/water-pollution-everything-you-need-know.

2 Hasegawa, Hiroshi, et al. *Environmental Remediation Technologies for Metal-Contaminated Soils.* Springer, 2015.

DEEP TROUBLE: PLASTIC POLLUTION IN OCEAN TRENCHES

Kyra Bauer

"The sea, once it casts its spell, holds one in its net of wonder forever."
— Jacques Yves Cousteau

Pollution in the ocean is a big problem, but it's not just the pollution floating on top or drifting just below the surface that threatens the sea. Pollution has been found almost seven miles down, in the deepest point on Earth—the Mariana Trench. We must address this problem before the damage becomes irreversible. Some actions have been taken to protect our planet's ocean trenches, but more needs to be done. Deep-sea pollution is working its way up the food chain straight toward humans.

Plastic is undoubtedly the worst offender when it comes to ocean pollution. About 80 percent of the trash drifting in the ocean is plastic. Roughly 8,800 tons (8 t) of plastic enters the oceans annually. That's like a garbage truck dumping a full load of plastic into the sea every minute.[4] It might be hard to believe, but even washing our clothes contributes to deep sea ocean pollution. Much of our clothing is constructed using fabrics made of various forms of plastic products. When these tiny bits, called microfibers, are shed from our clothing during washing, they end up in waterways that often lead to the ocean According to a report presented at the 2017 United Nations Marine Conference, as many as 51 trillion microplastic particles—500 times more than the stars in our galaxy—litter our oceans and seas, seriously threatening marine wildlife.[5] Some of these particles—as well as larger pieces of trash—regularly make their way to the deepest point on the planet, where a unique and mysterious **ecosystem** exists.

Figure 4 The lizardfish inhabits the deep sea.

Earth's deep-sea trenches bubble up superheated liquid **sulfur** and carbon dioxide. Active volcanoes relieve some of the Earth's pressure deep down. Trench ecosystems house marine life that has **adapted** to pressures 1,000 times that at sea level. As pollution makes it way down to the trenches, marine life must face disruptions in their fragile existence.

Even the smallest creatures in the sea are ingesting plastic and microfibers. "A British research team captured **amphipods**, tiny shrimp-like crustaceans that scavenge on the seabed, from six of the world's deepest ocean trenches and took them back to their lab. There, they discovered that more than 80 percent of the amphipods had plastic fibers and particles in their digestive systems."[1]

Microfibers are one of the most **insidious** pollutants because we can't readily see them, and therefore we don't think about the trouble they cause. When we wash our clothing in the washing machine, little fibers come off into the wash water. This is similar to when fibers come off our clothing in the dryer and collect in the lint trap. In the washing machine, these fibers leave our homes with water from our washers and travel to nearby wastewater treatment plants. There, many of them will settle into the sewage, but some will remain in the final treated wastewater that is released directly into our waterways that eventually travel to the oceans. Although washing our clothes in washing machines is just one source of microfibers in the environment, we know that it's a significant source. Certain fabrics release more microfibers than others—and some release none—but we know that **synthetic** fleece, in particular, sheds millions of microfibers that get into the oceans where amphipods eat them—and die as a result.

Figure 5 Plastic pellets known as nurdles

Microfibers aren't the only thing affecting life in the deep-sea trenches. Much of the plastic pollution that ends up in the ocean is industrial waste called microplastic. These tiny bits of plastic debris are the result of the disposal and breakage of consumer products and industrial waste. Anything made of plastic will have some amount of waste, and often that waste is simply dumped into the ocean. About half of all microplastics are intentionally produced as tiny pellets, commonly known as nurdles. These are "very tiny pieces of manufactured polyethylene plastic that are added as exfoliants to health and beauty products, such as some cleansers and toothpastes. These tiny particles easily pass through water filtration systems and end up in the ocean and Great Lakes, posing a potential threat to **aquatic** life." President Barack Obama signed the Microbead-Free Waters Act of 2015, banning plastic microbeads in cosmetics and personal care products in the U.S.[3] However, nurdles are also considered a pre-product, which means they are used in the manufacture of larger plastic products. Trillions of nurdles are produced by manufacturers around the world, particularly China, annually. Transport of nurdles regularly leads to spillage directly into the ocean. Since microplastics look like food to a variety of marine animals, from **crustaceans** and fish to whales, they pose a serious threat to the health of the ocean's ecosystems, from the surface of the sea to the deepest ocean trenches, where microplastics ultimately settle.

Figure 6 Trash covers areas of seabed.

It's not just plastic and other trash, but also oil and chemicals bringing their own set of problems to the table. "Persistent organic pollutants, known as POPs, include **PCBs** and polybrominated diphenyl ethers (PBDEs), which can be used as electrical insulators and flame retardants."[2] Researchers have found these chemicals on the ocean floor. They were banned in the 1970s because of their toxicity, yet roughly 1.4 million tons (~1.3 million t) of these chemicals were produced from the 1930s onward. "Industrial accidents, leaks from landfill sites and deliberate discharges saw large amounts of the chemicals get into the sea, which has often been used as a waste dumping ground."[2] These chemicals do not easily break down. At the bottom of the ocean, they continually affect lifeforms.

Pollution in the deep ocean touches every point in the **food web**. Dr. Susanne Brander, an Oregon State University toxicologist who researches the effects of microplastics on **larval** fish, says "the amphipods are becoming a vector to transport plastic particles into the food web." She explains that the amphipods that have ingested microfibers are "**prey** for larger fish, and those larger fish are prey for even larger **predators**. This puts the microfibers into the food web. We are finding larger organisms with intestines lined with microfibers. They found a baleen whale that had been beached and when they cut it open, the intestines were lined with these smaller particles. They give us a snapshot of what's happening in a larger sense."[1]

Figure 7 A deep-sea creature commonly called a sea pig

Humans may not live in the ocean and certainly don't ingest plastic pellets, but plastic pollution is making its way into our lives. Dr. Brander explains that "there are as many as 51 trillion pieces of plastic in the ocean, and 90 percent of that ocean plastic is microscopic. Scientists have previously likened it to a kind of 'soup.'"[1] It's almost impossible to get rid of the chemicals and garbage that are currently polluting the Mariana Trench and other areas of the seabed, but there are actions we can take to reduce the amount of new pollution that sinks to the ocean floor.

A first step is for each person to simply recognize his or her plastic footprint. How many wrinkled plastic bags do we shove into the trash every day? In America, the average household goes through 1,500 plastic shopping bags per year. How many single-use plastic cups do we use each year? Roughly 40 billion take-out utensils are produced a year and over 500 million single-use coffee cups are added to our landfills annually. And don't forget beverage bottles and caps, wrapping, straws, and plastic pens and pencils, as well as the myriad disposable containers and packages for food and other products. The average American uses 220 pounds (100 kg) of plastic annually.

Figure 8 A deep-sea creature called a chimaera

Regulations and laws that reduce industrial waste and pollution are necessary for the future of our planet. But such responses are slow and face many obstacles from industry. As individuals, we can start saving the planet right now. Reducing our reliance on plastic is a first step that we can all take toward protecting our deep-sea trenches and healing our oceans. Humans are the primary cause of plastic pollution in the ocean, and if we don't do something soon it might be too late to reverse the damage.

---

1 Gibbins, Sarah, and Laura Parker. "Creatures in the Deepest Trenches of the Sea Are Eating Plastic." Deep Sea Creatures in the Mariana Trench Eat Plastic, Study Finds, *National Geographic* 28 February 2019, www.nationalgeographic.com/environment/2019/02/deep-seacreatures-Mariana-trench-eat-plastic/.

2 Johnston, Ian. "Scientists Discover Banned Toxic Chemicals in Animals Living in World's Deepest Ocean Trench." *The Independent*, 14 February 2017, www.independent.co.uk/environment/toxic-chemical-levels-marianas-trenchanimals-pollution-sea-life-fish-worlds-deepest-pacific-ocean-a7578026.html.

3 National Oceanic and Atmospheric Administration (NOAA). "What are Microplastics?" *NOAA's National Ocean Service*, 13 April 2016, oceanservice.noaa.gov/facts/microplastics.html.

4 "Plastic Pollution – Sea Save Foundation." *Sea Save Foundation*, seasave.org/plastic-pollution.

5 Solheim, Erik. "The Missing Science: Could Our Addiction to Plastic be Poisoning Us?" *United Nations Environment Programme*, 4 June 2018, www.unenvironment.org/news-and-stories/story/missing-science-could-our-addiction-plastic-be-poisoning-us. 2015.

SPECIES AT RISK: AUSTRALIA'S GREATER BILBY

Kylie Davis

"If we save our wild places, we will ultimately save ourselves."

— Steve Irwin, the Crocodile Hunter

Australia's outback is the home to many unusual species that are not found anywhere else. In addition to the well-known koala, kangaroo, Tasmanian devil, and wallaby, the outback is home to a small **marsupial** that is a very important member of its ecosystem: the greater bilby. This animal is about the size of a rabbit. It has a long, pointed nose, pale blue-grey fur with a tan belly, big ears, and a crested black and white tail. It is **omnivorous**, feeding on everything from seeds, fungi, and plants, to small mammals and lizards. Australia once had two bilby subspecies: the lesser bilby and greater bilby. The lesser bilby was driven to **extinction** in the 1960s by feral cats and invasive rabbits. Now the greater bilby is similarly threatened with extinction.

The greater bilby inhabits deserts, dunes, and grasslands. As a relative of the bandicoot, the greater bilby is also known as the rabbit-eared bandicoot. Large, hairless ears help the bilby listen for predators. Greater bilbies are **nocturnal**. Their eyes are particularly sensitive to sunlight. They spend much of their time underground. Bilbies have strong jaws with 48 teeth suited to their diet of armored insects and tough plants. Powerful front paws help them dig, transforming their environment. Within their **territory**, they may dig as many as 12 holes that are up to 9 feet (2.7 m) deep. Plant material falls through the holes and decomposes, allowing new vegetation to grow around the surrounding land. The holes also become homes for other species such as the brush-tailed mulgara and spinifex hopping mice.

Figure 9 Greater bilby

The greater bilby once occupied 70 percent of Australia's landscape. With the introduction of invasive species, and native species negatively impacting the greater bilby population, greater bilbies are now in **fragmented** habitats mostly in western Australia near Broome through the Tanami and Great Sandy Deserts, and in scattered parts of southwestern Queensland.

Fewer than 10,000 greater bilbies exist in the wild. The Queensland Government Department of Environment and Science reported that "in Queensland, the impact of predators such as foxes and **feral** cats has had the greatest effect on greater bilby numbers and continues to be the most serious problem."[1] Australia's outback is flooded with feral cats. Data from nearly 100 wildlife surveys showed that there are between two and six million cats occupying the continent.[2] These feral cats were brought to Australia from Europe. Australia and Antarctica are the only continents to have no native cats, but the invasive feral cat population has exploded in Australia, which has spelled disaster for bilbies. Australia's "native wildlife **evolved** without the necessary defenses against felines, making them extremely vulnerable to predation by invasive cats."[2]

Feral cats are not the only threat to bilbies. The small marsupials must compete for resources with an invasive rabbit species also brought to Australia from Europe 200 years ago. The greater bilby and the rabbits fight over the land on which they both graze. Rabbits take over the many holes that bilbies dig—holes that other species rely on as well. Rabbit Free Australia notes that rabbits "are a significant contributor to widespread soil **erosion**

Figure 10 Haigh's Chocolate Easter Bilby

(impacting infrastructure and waters), and the destruction of perennial pasture plants, particularly when their populations are high."[3] These rabbits do the opposite of what bilbies do in the environment. While bilbies contribute to the system of restoration of soil and regrowth of vegetation in their habitats, rabbits destroy the soil and vegetation system. In addition, rabbit-caused soil destruction changes flood patterns and may cause runoff that could further damage the landscape.

Australia is committed to addressing the plight of the greater bilby. Australia is working to manage the problems of feral cats and invasive rabbits with the use of a variety of strategies. While not a permanent solution, sanctuaries protect bilbies from predators. Some sanctuaries provide secure breeding grounds in an attempt to revive the declining bilby population.

The Queensland government, Bush Heritage Australia, Save the Bilby Fund, Taronga Sanctuary, Rabbit Free Australia (RFA), and Zoos South Australia among others operate programs for the awareness, preservation, and protection of greater bilbies. The Queensland government has called on the public to assist in the monitoring of bilby populations by reporting bilby sightings in the wild. Social media has contributed to this plan.

Figure 11 Greater bilby

Another way Australians are working to help the greater bilby is by promoting the Easter Bilby (as an Easter Bunny replacement). Chocolate companies such as Organic Easter Bilbies and Haigh's Chocolate have raised public awareness of this **endangered** species. Funds raised from the sale of chocolate bilbies are donated to bilby conservation programs. The fight to save the greater bilby is ongoing. But with our understanding and effort, these unique marsupials just may stand a chance at survival.

---

1 "Greater Bilby." *Queensland Government Department of Environment and Science*, 2 July 2019, environment.des.qld.gov.au/wildlife/threatened-species/endangered/bilby#toc-0.

2 Grundhauser, Eric. "Nearly 100% of Australia Now Has a Feral Cat Problem." *Atlas Obscura*, 5 Jan. 2017, www.atlasobscura.com/articles/nearly-100-of-australia-now-has-a-feral-cat-problem.

3 "The Rabbit Problem." Rabbit Free Australia, www.rabbitfreeaustralia.com.au/rabbits/the-rabbit-problem.

THE GREAT BARRIER REEF: EXTINCTION OF AN ECOSYSTEM?

Jonathan Dodson

"The most single revelatory three minutes for me was the first time I put on scuba gear and dived on a coral reef."

— David Attenborough

The Great Barrier Reef, located famously off of Australia's Gold Coast, is world renowned and makes up about 10 percent of the world's coral reef ecosystem. The Great Barrier Reef has been referred to as one of the most complex natural systems and is one of the natural wonders of the world. This natural marvel draws tourists and scientists alike to wonder at the amazing underwater world and study the ecosystem. Unfortunately, we are only 33 years away from the Great Barrier Reef's mass destruction.[2] Finding a way to save the Great Barrier Reef and other reefs around the globe is essential for the well-being and survival of the planet.

There is a conflict surrounding the Great Barrier Reef. The conflict is between the preservation of the ecosystem and the economic impact of tourism. The Great Barrier Reef brings in an abundance of tourism resulting over $5.6 billion each year.[4] Should this reef cease to exist, the impact to the Australian economy and the general livelihood of those who rely on the reef would be substantial. In addition to the commerce that the reef provides, the ocean provides a barrier of protection, clean air, an entire food web, and ways to obtain medicinal ingredients/properties. The Great Barrier Reef is itself a wonder, but it is a wonder we truly cannot afford to lose. Without significant changes and help supporting the reef, we will surely see the repercussions of the damage we have caused—and the loss could be unrecoverable.

Figure 12 The Great Barrier Reef is home to a great diversity of life.

It is important to understand the nature of a reef system and the delicacy of the ecosystems. Coral reefs are large underwater structures composed of skeletons from marine **invertebrates** called coral. These reefs play a vital role in the protection of the shore. Coral reef systems have developed over a considerable period of time. Most of the substantial coral reefs found today range in existence from 5,000 to 10,000 years old. On the other hand, it would take only a fraction of that time for the reef to face serious problems that lead to its demise. For example, it takes a persistent water temperature increase of only 1.8 °F (1 °C) above average to disrupt the ecosystem. Once the damage to a reef system is done, it is an almost impossible process to ensure that life returns. This is a process new to scientists as they are still learning and applying measures to try to save multiple reefs around the ocean.

It is critical that we act now to stop reef deterioration. Coral reefs are vital ecosystems, but they are also rare and fragile. Coral reefs cover less than 1 percent of the ocean floor but account for more than 25 percent of all marine life. The Great Barrier Reef itself spans 344,400 square miles (891,992 sq km) and consists of about 3,000 individual reefs within it. These reefs are home to 1,625 types of fish, 30 species of whales and dolphins, 133 varieties of sharks, rays, 600 types of soft and hard corals, and thousands of other invertebrate species. In addition, the reef impacts 600 islands.

Figure 13 The Great Barrier Reef is a fragile ecosystem.

While the reef itself clearly provides for a vast part of the oceanic environment, the ocean also provides for the human population. The Great Barrier Reef supports the human population in several ways. Coral reefs help improve air quality, control erosion, and provide an economic boost. The annual revenue and ongoing job creation from the tourism industry is a huge component of the Australian economy, and it is all intertwined with The Great Barrier Reef. In fact, the reef supports 64,000 jobs in Australia and is a multi-billion-dollar contributor to the Australian gross domestic product.[7]

Unfortunately, while the reef has provided for the human population in many ways, we have not been so kind in return. **Climate change**, pollution, and tourism have led to the reef losing half its cover. Experts estimate that roughly 17.6 billion pounds (~8 billion kg) of plastic pollution enters the ocean annually. Plastic poses threats not only as trash but also as microplastics and chemical dispersion. Researchers estimate that by the year 2050, plastic will outweigh all the fish in the world's oceans.[5] Most plastic finds its way to the ocean floor—even down to the deepest trenches—where it is beyond our reach for retrieval, and much pollution has contaminated the Great Barrier Reef. Overfishing of its main predator, the giant triton sea snail, has caused the population of crown-of-thorns sea stars to unnaturally explode, resulting in devastation of entire areas of reef as thousands of the native starfish eat through it.

Figure 14 Without predators, the crown-of-thorns sea star is not being kept in check.

Climate change is also having devastating effects on reefs around the world. The Great Barrier Reef is a prime example of what happens when the temperature and gas changes in the atmosphere even a fraction of what the ecosystem is accustomed to in nature. When temperatures rise or lower just slightly, the oceanic environment begins to see harsh repercussions. One effect of climate change is the process of coral bleaching. The process entails coral moving from a healthy state to a vulnerable or bleached state. In the healthy state, the coral has a **symbiotic** relationship with an algae food source. Increased pollution or temperature moves the algae to a stressed state in which the algae begins to leave the coral until ultimately the coral reaches a bleached status. The bleaching of coral leaves it susceptible to disease that then leads to its overall deteriorating of the ecosystem.

It is clear that the deterioration of the Great Barrier Reef is a known issue and, while it has been monitored, little has been done to stop the process. Every five years the Australian government issues an outlook report. The 2019 report confirms the greatest threat to the reef still remains the changing of the Earth's climate.[6] The reef has seen multiple major bleaching events in just the past two decades as a direct result of climate change. Each bleaching event comes more frequently compared to previous events, and the outcomes are becoming much harsher with each instance. Other threats include coastal development, land-based waste run-off, and human use.

Figure 15 Coral bleaching leaves only skeletal remains of once-living coral communities.

Because it causes significant damage, coral bleaching has been studied for some time. Researchers know the impact and the timeline of this process. Terry Hughes, a marine biologist at James Cook University and one of the world's top reef researchers, has explained that bleached coral takes about 10 years for the fastest growing corals to have a chance to recover from a bleaching event; however, the problems with the 10-year window are that the chances of a new bleaching event occurring are very high due to global warming/climate change.[1] It is clear that if we do not acknowledge climate change and the harsh effects it has on ecosystems—and make immediate changes to our actions as a society—we will end up seeing swift damage upon ecosystems that took thousands of years to form.

Considering the proposed 30 years before mass destruction, it is clear that actions need to be taken to ensure the safety of not only reef ecosystems but also coastal ecosystems. The oceanic environment, and coral reefs more specifically, are intertwined with at least 14 coastal ecosystems.[3] While we have seen slight improvements in protective policies and an acknowledgment of the dangers of **greenhouse gases**, we have not made significant and lasting changes. Climate change and tourism are not the only factors harming the reef ecosystems. Measures need to be undertaken for the protection of the Great Barrier Reef in terms of pollution, overfishing, and the industrialization of the environment. Australia has put in place a 50-year plan to try to help preserve the Great Barrier Reef. While this is a step in the right direction, this plan alone will not stop the destruction of reef systems, and preservation does not stand a chance without worldwide action and acknowledgment of the issue.

Figure 16 The Great Barrier Reef draws more than a million visitors annually.

Education is one of the best forms of prevention and conservation action. Education on the effects of climate change on a more elementary level is needed. Even those presently in Australia do not seem to grasp the totality of what they could face if no change is implemented. This topic needs to have a light shed on it. One of the best things we can do is to create an informational program to help youth and adults understand what is happening and the long-term impacts. Humans seem to not acknowledge the damage until it is too late. The mass bleaching of the reef systems has been compared to cutting down the redwood trees that are in some cases thousands of years old. This example alone may be enough for some people to recognize the devastation we are causing. The policies for education worldwide to share the impact of human actions would help but only to the extent that action followed—policies and actions performed at a higher level are also needed.

Another way for Australia to help the reef would be to limit the industrialization and tourism within the country and surrounding islands. While this may have economic impacts, it would make a difference in the pollution volume, temperature increases, and overall industrialization of the environment. Additionally, Australia could put protective procedures in place for species most at risk, such as corals, rays, and sharks. Even if these measures were seasonal, it may give the environment time to adapt. Any changes that are implemented must be considered long-term commitments.

Figure 17 Coral reefs are like vast underwater forests.

The situation in Australia surrounding the rapid decline of the Great Barrier Reef needs attention and action. If we do not make a significant change to the factors that are negatively affecting the reef, it may not last our lifetime. In addition to the significant cultural loss, the physical changes and impact of the loss of an entire ecosystem may have far-reaching impacts that are not yet predictable. As species become extinct and life ceases to exist in the ocean surrounding the Gold Coast, humans must be held accountable for the irrevocable damage that they are causing. While it is not too late to change our path and create a **sustainable** environment, every day we are letting our opportunity slip away. The human race needs to ensure the reef is maintained for the future of marine life and the humans that study and visit the ocean. While it may seem that an individual has little power to impact this world issue, one step everyone can take is simply sharing an article about coral reefs on social media for others to see and hopefully further spread knowledge by participating in a chain of education for implementing change. Those with knowledge need to continue to challenge others and share their awareness. Leo Tolstoy once said, "Everyone thinks of changing the world, but no one thinks of changing himself." By changing oneself and the habits we have, we can change the world and save the Great Barrier Reef.

---

1 ABC News (Australia). "The Reef Pt 1: Is it too Late to Repair the Great Barrier Reef?" *YouTube*, 5 April 2020, youtu.be/Rmkyj9qghGY.

2 ABC Science. "Can We Save the Reef?" *YouTube*, 6 April 2020, youtu.be/nYXFEfB2Lf8.

3 "Coastal Ecosystems." *Great Barrier Reef Marine Park Authority*, 2020, gbrmpa.gov.au/the-reef/coastal-ecosystems.

4 "Great Barrier Reef Supports 64,000 Jobs and Generates $6.4 Billion for Australian Economy, Deloitte Calculates." *National Geographic*, 26 June 2017, blog.nationalgeographic.org/2017/06/26/great-barrier-reef- supports-64000-jobs-and-generates-6-4-billion-for-australian-economydeloitte-calculates/.

5 "Ocean Pollution: 11 Facts You Need to Know." *Conservation International*, 2020, conservation.org/stories/ocean-pollution-11-facts-you-need-to know.

6 "Outlook Report 2019." *Great Barrier Reef Marine Park Authority*, 2020, gbrmpa.gov.au/our-work/outlook-report-2019.

LEATHERBACK SEA TURTLES: GENTLE OCEAN GIANTS

Abigail Dye

"We can share beaches and oceans with sea turtles, but it requires commitment and effort on our part."

— Victoria B. Van Meter, *Florida's Sea Turtles*

Up to eighty percent of the planet is affected by ocean pollution. Sarah Nelms, et al. wrote this about ocean pollution: "Its persistence and dispersal throughout marine ecosystems has meant that sensitivity toward the scale of threat is growing, particularly for species of conservation concern, such as marine turtles."[6] One such species affected by ocean pollution is the heaviest **reptile** on Earth: the leatherback sea turtle.

Seven species of sea turtle inhabit the world's oceans. All but one species has a hard, domed **carapace**. The leatherback sea turtle instead has somewhat flexible and almost rubbery skin on its back. This leaves leatherbacks vulnerable to injuries and attacks by predatory sharks. But one threat that affects leatherbacks more than anything else can be blamed entirely on humans. Ocean pollution kills countless sea turtles year after year.

All kinds of waste such as oil, chemicals, trash, and microplastics comprise ocean pollution, which results in damage to the environment, to the health of all organisms, and to economic structures worldwide. One major type of ocean pollution is runoff. The National Oceanic and Atmospheric Administration (NOAA) has reported that "nonpoint source pollution includes many small sources, like septic tanks, cars, trucks, and boats, plus larger sources, such as farms, ranches, and forest areas."[5]

Figure 18 Females leatherback sea turtles come ashore to lay eggs.

Individuals and not only industry are often responsible for runoff that ends up in the ocean. Failing to properly dispose of oil, chemicals, paint, and other such substances can allow these toxins to end up in storm drains and water systems that empty into streams, rivers, and other waterways that eventually carry the pollution to the ocean.

Adult leatherbacks feed almost exclusively on jellies that drift with the currents. For this reason, plastic pollution—particularly bags that drift in the water and can be mistaken for their favorite meal—end up being acutely deadly for leatherback sea turtles. "Sharp plastics can rupture internal organs and bags can cause intestinal blockages, leaving turtles unable to feed, resulting in starvation."[7]

Baby sea turtles are also at risk. When hatchling turtles emerge from their eggs in nests laid on dry land and then make their way back to the water, they may be forced to shuffle through plastic waste, risking entanglement. Only 1 out of 1,000 leatherback hatchlings make it to adulthood. While plastic is not the main cause of their demise, it does add to the peril they face.

Why should we care about leatherback sea turtles? Because they are beneficial to marine ecosystems and by extension to humans. As members of a massive food web, they participate in maintaining the ocean's health. For example, they feed on jellies, which must be kept in check. Rising ocean temperatures are leading to an explosion of jellies. Although jellies may seem like small links in the food chain, "in large **aggregations**, jellyfish have damaged the economic success of power plants and fisheries. They consume larvae of commercial fish species and prevent the recovery of overfished populations."[4] As jelly populations increase, other species are **displaced**. Without leatherbacks feeding on them, jellies could more easily overpopulate.

Figure 19 Leatherback sea turtle hatchlings

Leatherbacks also help beaches. Unhatched turtle eggs and undeveloped hatchlings provide nutrients for beach vegetation, which stabilizes dunes and helps to prevent coastal erosion. As part of the food web, hatchlings that do not make it to the sea feed a variety of invertebrates, birds, and small mammals. A decrease in the leatherback sea turtle population will impact many other animals.

Threats to sea turtle populations are serious, and there are steps we can take to address the leatherback's plight. We can clean up our oceans, properly dispose of trash, and become educated on the importance of protecting the ocean and its inhabitants. We can recycle trash, stop relying on single-use plastics, and start incorporating materials that are reusable into our daily lives. Many organizations are doing this and encouraging others to join in, and people are noticing.

Some ocean cleanup organizations are Greenpeace, Oceana, Take 3, and 4Ocean. These organizations are raising awareness of the importance of cleaning up our oceans. Some are even making ocean clean-up a fun activity for their volunteers and for everyone around the world. Learning about these organizations and their goals can help people understand the problem while suggesting solutions. Organizations are promoting their works primarily through social media and print and video sources. Many celebrities use their platforms to encourage people to pitch in. For example, actor Leonardo DiCaprio is a well-known environmentalist and philanthropist who promotes education about climate change and the need to protect and preserve the planet's resources. The Leonardo DiCaprio Foundation has donated tens of millions of dollars to a variety of organizations for environment protection.

Greenpeace is a global, independent campaigning organization that uses peaceful protest and creative communication to expose global environmental problems and promote solutions that are essential to a green and peaceful future. Greenpeace has offices all over the world, including all U.S. states. Of Greenpeace, the organization's executive director, said, "We 'bear witness' to environmental destruction in a peaceful, non-violent manner. We use non-violent confrontation to raise the level and quality of public debate. In exposing threats to the environment and finding solutions we have no permanent allies or adversaries. We ensure our financial independence from political or commercial interests."[1]

Oceana is the largest international advocacy organization focused solely on ocean conservation. Their goal is to make our oceans as healthy and abundant as they once were. They have many campaigns that explain the issue of ocean pollution and have done plenty to try to help. For example, they had a campaign to show how marine life feeds many families, and every dollar that was donated was used to feed these families.

Take 3 is an organization that believes humans can correct problems with ocean pollution by taking simple actions. The Take3 for the Sea goal is "to conserve the environment and protect wildlife from the impacts of plastic pollutions and waste by leading a movement of people connected to the planet."[3] The plight of sea turtles inspired this organization. Take 3 is an organization that encourages people to take at least three pieces of trash they find on beaches and throw them away or recycle them.

4Ocean is one another major ocean cleanup organization. Its goal is to get as many people as possible involved in ocean cleanups. It sells bracelets made mostly from recycled materials as well as merchandise such as apparel made of materials that have been environmentally and socially responsibly sourced. Every bracelet that is bought prevents one pound of trash from ending up in the oceans.

These organizations are important to cleaning up ocean pollution, but individuals are needed as well. Throwing away trash without overflowing trash cans as well as recycling and utilizing materials other than plastic are just a few of the simple steps that everyone can take to contribute to protecting the ocean from pollution.

Figure 20 It is important to keep our beaches clean.

Brian Hutchinson at Ocean Society says, "With skyrocketing plastic production, low levels of recycling, and poor waste management, between four and twelve million metric tons of plastic enter the ocean each year." This amount is estimated to double over the next ten years.

By reducing the production and consumption of single use plastics, we can cut back greatly on plastic trash. Single use plastics include such items as straws, shopping bags, drink bottles, packaging, plates, cups, and utensils. Some businesses are taking charge and getting a jump start on using less plastic in their facilities. Even businesses located far from the ocean realize the importance of reducing plastic waste that could eventually make its way to the sea. For example, the Fireside Grille in Winfield, West Virginia, places signs on their tables that inform patrons of the restaurant's commitment to doing its part to protect marine life by encouraging diners to choose a paper straw or no straw at all. Because of this strategy, returning patrons have been spotted in the restaurant bringing reusable straws from home.

Reducing the amount of plastic that we use every day is a great start to addressing the problem of ocean pollution. And it's important to remember that many types of plastic can be recycled. However, most people find it difficult to completely eliminate plastic from their lives. It may be helpful to consider that other materials can be used to replace much of the plastic in our everyday lives. **Bioplastic** products made of such natural substances as sawdust, cornstarch, rice, and hemp are a good alternative. Bioplastics are ecofriendly because they are biodegradable, meaning they break down harmlessly in the environment.

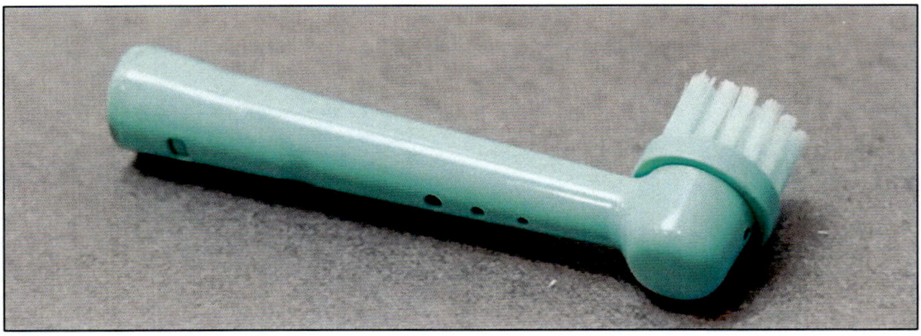

Figure 21 TIO was founded in 2015 to produce bioplastic electric toothbrush heads.

TIO is a company based in Berlin, Germany, that manufactures electric toothbrush heads from sustainably farmed sugar cane. The process involved in TIO's manufacturing of toothbrush heads is outlined on their website: tio.care/nachhaltigkeit. Of course, more familiar cardboard and paper could be used more often for packaging and then recycled. In grocery stores, there is a good chance that the bags are already recycled and can be reused over and over again. If cardboard and paper absolutely have to be thrown away, at least they break down naturally and much faster than plastic. Also, a lot less chemical residue is left after cardboard and paper break down.

Trees are not an infinite resource, however. This is why stones are used to make paper products as well. Stone-plastic and stone-paper are made from calcium carbonate and can replace shopping bags, food cartons, and other containers and storage bags. Milk can also be used as a plastic replacement. Milk-plastic is made from casein, a protein found in milk. The casein is mixed with clay and a reactive molecule that makes it sturdy. The packaging is a harmless way to reduce plastic waste. Milk-plastic is even edible, which means it can be fed to livestock.

Many plastics, especially lightweight items such as straws, wrappers, and shopping bags are blown and swept into waterways, with many eventually making their way to the sea. Ninety percent of marine waste is plastic. Wildlife can be suffocated, strangled, or poisoned. Plastic itself is toxic to animals, and the water can become contaminated from toxic chemicals leeching out of plastic trash. Fish that humans consume may contain toxins, which means humans could be eating the plastic trash we discard. The effects of plastic consumption on the human body are not fully understood. However, consider that since we know plastic pollution kills marine life, we could easily understand how it could be harmful to human bodies as well.

Leatherback sea turtles as well as the wildlife sharing turtles' ocean habitats struggle with a deadly host of human-made dangers posed by plastic pollution such as entanglement resulting in injury or drowning, plastic ingestion resulting in poisoning or starvation due to gastrointestinal blockage, and any number of other tragic consequences.

Figure 22 Sea turtle hatchlings face a perilous journey from the beach to the sea.

Plastic is convenient and easy to use, but it is destroying our global ecosystem and negatively impacting marine life. Leatherback sea turtles and their relatives are seriously at risk because of plastic. Wildlife is helpless when it comes to ocean pollution. But humans have the power and ability to help stop the harm caused by our own hands. Being mindful of what we throw away and understanding where it goes is a first step in taking control of our plastic habits. We can start or increase recycling and choose materials other than plastic. Ocean destruction does not have to be irreversible. We can make a difference starting today.

Figure 23 Leatherback sea turtles need our help.

1 "About." *Greenpeace USA*, www.greenpeace.org/usa/about/.

2 Hutchinson, Brian. "7 Ways to Reduce Ocean Plastic Pollution Today." *Ocean Society*. www.oceanicsociety.org/blog/1720/7-ways-to-reduce-ocean-plastic-pollution-today.

3 "Learn More About Take 3 for the Sea and Our Aims." *Take 3 for the Sea*, www.take3.org/about/.

4 Lee, Joong. "Jellyfish Apocalypse: Problems, Causes, and Opportunities." *Science Buzz*, 2018, www.sciencebuzz.com/jellyfish-apocalypse-problems-causes-and-opportunities.

5 National Oceanic and Atmospheric Administration (NOAA). "What Is the Biggest Source of Pollution in the Ocean?" *NOAA's National Ocean Service*, 8 October 2008, oceanservice.noaa.gov/facts/pollution.html.

6 Nelms, Sarah E., et al. "Plastic and Marine Turtles: A Review and Call for Research." *ICES Journal of Marine Sciences*, vol. 73, no. 2, 9 October 2015, pp. 165-181. *Oxford University Press*, academic.oup.com/icesjms/article/73/2/165/2614204.

7 "What Do Sea Turtles Eat? Unfortunately, Plastic Bags." *World Wildlife Fund (WWF)*, 11 October 2018, www.wwf.org.au/news/blogs/what-do-sea-turtles-eat-unfortunately-plastic-bags. 2015.

AFRICAN ELEPHANTS AND THE IVORY TRADE

Bradley F. Hall

"There is no creature among all the Beasts of the world which hath so great and ample demonstration of the power and wisdom of almighty God as the Elephant."

— Edward Topsell, *The Historie of Foure-Footed Beastes*

Two distinct elephant species live on the continent of Africa: the bush elephant (*Loxodonta africana*) and the forest elephant (*Loxodonta cyclotis*). Male African bush elephants can grow to be 10 to 13 feet (3–4 m) high at the shoulder and can weigh up to 18,000 pounds (8,165 kg). Forest elephants are slightly smaller and stouter than bush elephants. Their tusks are straighter and slenderer than the tusks of bush elephants. In both species, female elephants (called cows) are about half the size of males (called bulls). Females run their herds. A herd may consist of 6 to 12 family members and may grow to include 20 to 40 elephants from other families. Males are typically loners or live in all-male herds, called bull bands, except during mating season. Elephants are essential to the health of many African ecosystems. Their **dung** fertilizes the grass and plants that they eat and spread seeds from the fruit and nuts they consume. They create watering holes for other wildlife. Elephants forage for food almost constantly—about 20 hours a day. These animals are vital components of the African food web.

For hundreds of years, elephants have been hunted for their meat, hide, bones, and ivory tusks. Both male and female elephants develop tusks. Tusks are actually long teeth that stick out when the elephant's mouth is closed. A male bush elephant's tusks may be nearly 8 feet (2.4 m) long and weigh up to 250 pounds (113 kg) each.

Figure 24 Only about 350,000 African elephants remain.

Since 1989, hunting of the African elephant and ivory trading have been illegal, but professional traders have been trying to get the laws overturned. The current laws have done little to stop individuals from participating in the illegal ivory trade. Since the ban went into effect, the value of elephant ivory has increased 1,000 percent. Elephant ivory is highly sought after for **trophy** collecting and folk medicine. Currently, a single pound of elephant ivory can fetch $1,500 US.

Conservationists and wildlife scientists have sought solutions to **poaching** and the ivory trade, but nothing has been able to stop it completely. Elephant body parts are used in folk medicine. It is believed that ivory can cure sore throats, indicate poisoned food by changing color upon contact, heals sores and boils, and beautify skin.[4] Such claims are myths, yet the demand for ivory persists, so poachers continue to slaughter elephants.

Many conservation organizations such as World Wildlife Fund (WWF) and the International Anti-Poaching Foundation (IAPF) support awareness and education about poaching by sponsoring events at schools, markets, youth camps, government meetings, and rural villages. WWF is among the many organizations that conduct elephant research. One of WWF's projects is the **satellite**-tracking of forest elephants in West Africa. The information gathered from **GPS** devices placed on individual elephants helps researchers count herd populations and understand how and where the elephants **migrate** seasonally. Learning about elephants is important to the development of strategies to protect them.

But poachers are persistent. They employ the use of **night vision** equipment; **tranquilizers**, poisons, and traps; and weapons with sound suppressors to sneak up on the elephants. When a poacher kills an elephant, the tusks are hacked out of the elephant's upper jaw. There is no way to steal an elephant's tusks without destroying the animal.

Elephants keep the grasslands and forests in balance. If elephants are driven to extinction, then more animals in the food web will be affected. However, nature sometimes finds a way to combat stresses when it can. Elephants seem to be adapting to ivory poaching in an interesting way: tusks are disappearing over generations of elephants. As this continues, poachers are being deprived of the prize they seek. The *Independent* reported that "in 2008, scientists found that even among elephants that remained tusked, the tusks were smaller than in elephants a century before—roughly half their previous size."[2] This suggests that nature is taking steps to protect elephants from poachers. CBS News recently reported that "researchers are still trying to pinpoint all the **genetics** that underpin tusklessness. However, poachers are essentially weeding out the gene for tusk growth from elephant populations."[3] By targeting elephants with the biggest tusks, poachers are actually eliminating the large-tusked elephants from the **gene pool**—thus eliminating the very resource they wish to exploit.

Figure 25 Female elephants are being found with smaller than normal tusks.

Dominique D'Emille Correia Gonçalves, a researcher who studied elephants at Gorongosa National Park in Mozambique, found that about one-third of females born in recent years did not have tusks, and many others had smaller than normal tusks. And in South Africa's Addo Elephant National Park, 98 percent of female elephants born since 2000 were tuskless.[1] When poachers kill elephants that carry this gene, the trait cannot be passed on, leaving only small or no-tusk genes to carry forward.

Wildlife veterinarian Michael D. Kock recently reported that among the elephants in North Luangwa, Zambia, roughly 60 percent were tuskless. Kock believes that the lack of tusks is directly related to poaching because among elephants in protected areas, where poaching was curbed, tusks reappeared among newborn elephants.[1] This suggests that the gene for large tusks can re-emerge if elephant populations increase.

The poaching of elephants in Africa is pushing these animals toward extinction and thus impacting their ecosystem. Poaching is a large-scale problem that it seems impossible to stop. Only when demand for ivory drops will poaching stop. With organizations continuing their many programs in education and research, and with stronger laws and stiffer penalties for those involved in the ivory trade, hopefully we can help African elephants survive.

Figure 26 African elephants need our help.

[1] Bentson, Clark. "Poaching Likely Created a Generation of Tuskless Elephants, Scientists Say." *ABC News*, 16 January 2019, abcnews.go.com/International/poaching-created-generation-tuskless-elephants-scientists/story?id=60388821.

[2] England, Charlotte. "Elephants Are Now Being Born without Tusks Because of Poaching." *The Independent*, Independent Digital News and Media, 27 Nov. 2016, www.independent.co.uk/news/elephants-africa-tusks-ivory-poaching-born-without-a7440706.html.

[3] O'Kane, Caitlin. "African Elephants Are Evolving without Tusks Because of Poaching." *CBS News*, CBS Interactive, 27 Mar. 2019, www.cbsnews.com/news/african-elephants-are-evolving-to-not-grow-tusks-because-of-poaching/.

[4] Wei, Clarissa. "The Myths of Medicinal Ivory." *KCET*, 4 Feb. 2016, www.kcet.org/food/the-myths-of-medicinal-ivory.

GETTING TO KNOW THE ARCTIC FOX

Jessica R. Jenkins

"One of the most adorable and loveable animals in the animal kingdom is the arctic fox."

— Louis Dorfman, Animal Behaviorist

Arctic foxes are found in the Arctic **tundra**, throughout Alaska, Canada, Greenland, Russia, Norway, Scandinavia, and Iceland. These foxes are known for their beautiful white fur, which, unfortunately, humans love to buy and wear. Arctic foxes are not always white. Their winter coat is white to blend in with the snow, but in summer they grow brown or gray coats. In their place in the middle of the food web, arctic foxes are vital members of their ecosystems. These ecosystems could become unbalanced if arctic foxes are not around to keep smaller prey animals such as lemmings and voles from overpopulating. Such a dire situation is certainly possible due to human interference. Not only are we hunting arctic foxes, but we are also failing to attack the problem of climate change, which negatively affects these animals.

The arctic fox is a magnificent creature that lives in very cold, snowy environments. It is about the size of a domestic cat, weighing up to 17 pounds (7.7 kg). Its body is equipped to survive the cold in many ways. Thick fur helps it survive in the winter by keeping it warm and by providing **camouflage** so that it can avoid predators. It has the warmest **pelt** of any Arctic animal and has been known to survive temperatures as low as –58 °F (–50 °C). It has fur on its paws that "gives the fox traction so it does not slide on the ice." Also, "because its legs are so short and its ears and nose are so small, there is less of the fox exposed to the cold."[1] When it sleeps, it curls up and wraps its tail around its face for added warmth.

Figure 27 In summer, the arctic fox has a dark coat.

Figure 28. In fall, the arctic fox begins to grow a white winter coat.

The arctic fox plays a significant role in helping to balance its ecosystem. Some animals or even plants will overpopulate if something isn't keeping it controlled. It will affect me and you too! How you ask? Well, if deer go extinct, that's one food source gone for large predators such as bears, wolves, and even us. Also, deer eat poison ivy. If they do not exist to eat it, it could overpopulate, force out other plants, and become invasive in our forest, fields and gardens. Our **edible** plant sources, along with necessary animals who feed on certain plants, could be gone. One broken link in the food chain affects everything!

Like every creature on Earth, arctic foxes are important members of their ecosystems. They help balance everything. If we lose arctic foxes, the natural balance will surely be disrupted. Smaller prey animals may overpopulate and damage the environment, and larger predators such as bears and wolves may go hungry without arctic foxes on which to prey.

Arctic foxes eat voles and lemmings—rodents that are known carriers of various diseases. If arctic foxes are gone, then the lemmings and voles will overpopulate the ecosystem, severely threatening other members of the food web with overpopulation and disease. Humans could also be impacted. Lemmings carry a disease called Tularemia, which can be passed to humans by insects on the lemmings or through the handling of infected animals. This disease "causes fever, muscle pain, and symptoms associated with the point of entry into the body."[3]

What are the threats to arctic foxes? One major issue is overhunting. This is when people hunt so many animals that the population cannot recover quickly enough to replenish the species in a given area. People have been overhunting arctic foxes for decades because of the desire for their beautiful winter pelts. Hunting of this magnificent creature was at an all-time high in the early 1900's. In 1928, with arctic foxes nearly extinct in Sweden, the nation imposed a hunting ban. In 1930, Norway followed; and in 1940, Finland also outlawed hunting. After all this time, their numbers have not improved. In Sweden, the Arctic fox is listed as endangered, and in Norway, they are critically endangered. They have been **extirpated** from Finland. Once numbering over 20,000 in Norway alone, arctic foxes have continued to disappear, with fewer than 60 adult individuals still alive throughout all of Scandinavia in the year 2000. Despite this, poaching persists.

Poaching is not the only thing arctic foxes must face. They are also affected by climate change. They are suited to a cold environment, but Earth's Arctic is warming up. As temperatures rise, tree lines will move farther north, luring wildlife such as the red fox farther north as well. "Red foxes are twice the size of arctic foxes and not only compete for prey but can take [arctic foxes'] dens by killing them or chasing them away."[2] Such competition for resources is another way that arctic fox populations can decline, leading their ecosystems to become unbalanced.

Figure 29 Climate change is affecting the snow and pack ice on which arctic foxes rely.

Many people have taken action to save arctic foxes. Organizations such as the National Wildlife Federation, World Wildlife Fund (WWF), and Wild Rescue and Rehabilitation provide educational materials to the public and support protection of and research on arctic foxes. A bold initiative was launched in 2000, when the Norwegian Environment Agency funded the Arctic Fox Captive Breeding Programme. Its goal is strengthening existing populations and reintroducing arctic foxes to areas from which they have been eliminated. It had a rocky start, but it has since made some headway in saving arctic foxes, introducing several hundred individuals into the wild.

Learning about wildlife such as arctic foxes is the first step in becoming aware of the problems and potential solutions associated with them. We need to support hunting bans, avoid buying animal products, and do our part to work toward climate change solutions. We as citizens can help arctic foxes thrive once again if we all pitch in!

---

1 "Arctic Fox - Vulpes lagopus - NatureWorks." *New Hampshire PBS*, nhpbs.org/natureworks/arcticfox.htm.

2 Pomeroy, Donna. "11 Arctic Species Affected by Climate Change." *WWF*, www.wwf.org.uk/updates/11-arctic-species-affected-climate-change.

3 "Tularemia." *Dictionary.com*, www.dictionary.com/browse/tularemia.

THE GROWING THREAT OF HURRICANES

Shayne Miller

"No one can prevent hurricanes, but prosperous communities are much better able to withstand them than poor ones."

— Robert Zubrin, American aerospace engineer

Typhoons, cyclones, and hurricanes all go under the scientific name of tropical cyclone—the general term for all circulating weather systems over tropical waters. Hurricanes are high-speed windstorms accompanied by torrential rains. They can be very different in terms of precipitation and intensity, but, in general, hurricanes are said to be the most violent storms on Earth. Let's take a look at these weather occurrences and consider how climate change is affecting them.

The National Aeronautics and Space Administration (NASA) tells us that "tropical cyclones are like giant engines that use warm, moist air as fuel. That is why they form only over warm ocean waters near the **equator**."[5] Hurricanes occur when a cold front meets a warm front. Water with a temperature above 80 °F (26.7 °C), specific atmospheric conditions and a little bit of spin are enough to create hurricanes that can span over 1,000 miles (1,609 km). They usually last for about a week, moving 10 to 20 miles (16–32 km) per hour over the open ocean. According to the United States Search and Rescue Task Force, on average, roughly six Atlantic hurricanes occur annually.[6] Hurricanes result in a lot of property damage, and take a heavy toll on human lives, wildlife, and the ecosystem. The continued stress on our planet due to climate change is only intensifying the threats posed by hurricanes, with an increase in the frequency and intensity of hurricanes around the world.

Figure 30 Isabel, a Category 5 hurricane, was the deadliest, costliest, and most intense hurricane in the 2003 Atlantic hurricane season.

Figure 31 Irma was a Category 5 hurricane that caused widespread destruction in 2017.

Some places in the world are more vulnerable to hurricanes than others. Indeed, places such as the Atlantic Coast, the Gulf of Mexico, and the Hawaiian Islands are susceptible to major storms. The Atlantic hurricane season starts on June 1 and ends on November 30. The Eastern Pacific hurricane season runs from May 15 to November 30. It is important to keep in mind that most hurricanes occur during the fall months. Hurricanes always have a name. The National Ocean Service tells us that "there is a strict procedure established by the World **Meteorological** Organization. For Atlantic hurricanes, there is a list of male and female names which are used on a six-year rotation. The only time that there is a change is if a storm is so deadly or costly that the future use of its name on a different storm would be inappropriate."[6] For example, the names Katrina and Ike will not be used again. There are also different types of hurricanes, called categories, which depend on 5 different factors: wind, pressure calculated in millibars, pressure in inches, storm surge, and damage.

Climate change is an important factor when considering hurricanes. NASA's definition of climate change says it is "a broad range of global phenomena created predominantly by burning **fossil fuels**, which add heat-trapping gases to Earth's atmosphere. These phenomena include the increased temperature trends described by global warming, but also encompass changes such as sea-level rise; ice mass loss in Greenland, Antarctica, the Arctic, and mountain glaciers worldwide; shifts in flower/plant blooming; and extreme weather events."[2]

Understanding Global Warming[3]

Global warming is only one aspect of climate change. Global warming refers to the rise in global temperatures due mainly to the increasing concentrations of greenhouse gases in our planet's atmosphere.

Greenhouse gases are gases in Earth's atmosphere that trap heat. They let sunlight pass through the atmosphere, but they prevent the heat that the sunlight brings from leaving the atmosphere. The main greenhouse gases are

- water vapor
- carbon dioxide
- methane
- ozone
- nitrous oxide
- chlorofluorocarbons

Greenhouse gases are a good thing. Without them, our planet would be too cold, and life as we know it would not exist. But there can be too much of a good thing. Scientists are worried that human activities are adding too much of these gases to the atmosphere.

Global warming and climate change increase the frequency of hurricanes we have each year as well as their power and the resulting damage they cause. Indeed, scientists have linked two different things that go together in this case. Because of human activities, such as urban expansion, growth of industry and commercial agriculture, new technologies, resource exploitation, and so on, the amount of greenhouse gases released into the atmosphere has gone up. We have all noticed that temperatures have been changing; in fact, we often hear in the media that temperatures are very high for the current season. Heat from the warming atmosphere raises the temperature of the sea surface, which makes it easier for the water to reach that critical 80-degree mark, and therefore improves the chances that a hurricane will form.

Additionally, there are multiple further consequences of climate change. Warmer temperatures combined with an increase in atmospheric moisture content result in increased rainfall during tropical storms. And more rainfall means more flooding inland. According to the National Oceanic and Atmospheric Administration (NOAA), modeling studies project that a 3.6 °F (2 °C) increase in global temperatures would result in a 10 to 15 percent increase in rainfall rates within about 62 miles (100 km) of a storm.[4] While hurricane winds are powerful and highly destructive, the rainfall that accompanies hurricanes can be equally catastrophic.

Figure 32 Flooding in Port Arthur, Texas, caused by Hurricane Harvey in 2017

The number of hurricanes that reach very intense levels (categories 4 and 5) will likely increase due to continued warming in this century. It will be easier to predict how global warming will increase than to predict how the frequency and intensity of hurricanes will change in the next few decades, but there is one thing we know for sure: climate change has an impact. We should all be concerned with this problem because we are the ones who will be affected in some way by hurricanes. But our minds can stay positive because we can also be the actors of a changing world.

Rising sea levels are also part of the equation. Sea levels have risen by nearly eight inches (20 cm) since the dawn of the Industrial Age, and the rise is accelerating in response to warming. The main reasons are that the ocean waters expand as they heat up. As the planet warms, glaciers and continental ice caps melt. The water changing from solid to liquid means that the temperatures of the water in our oceans are going up; it also means that there is more water.

Places such as Greenland and Antarctica are melting at a higher pace than was expected in the 20th century because of accelerated climate change. Scientists have reported that Greenland holds enough ice to raise global sea levels by 23 feet (7 m) and Antarctica by 190 feet (58 m).[10] Rising sea levels mean greater flood risk. Coastal cities could be wiped out. Right now, more than 200 million people worldwide live along coastlines 16 feet (5 m) or less above sea level. By the end of the 21st century this figure is estimated to increase to 400 to 500 million people.[8]

Storm Surge[11]
Storm surge is the abnormal rise in seawater level during a storm, measured as the height of the water above the normal predicted astronomical tide. The surge is caused primarily by a storm's winds pushing water onshore.

The second ingredient in increased flood risk is changes in the properties of storms that cause storm surges, which can raise water levels 20 feet (6 m) or more above sea level. Sea level rise and storm surge are linked. Indeed, as a result of sea level rise, storm surges occurring today are eight inches (20 cm) higher than they would have been in 1900, for example. This number will keep increasing because it is linked to the number of hurricanes that occur each year and rising sea levels.

Water weighs about 1,700 pounds per cubic yard (1,000 kg per cubic m), so when that much water gets going, it can damage or destroy structures not built to withstand such forces. Impacts of storm surge include extensive property loss, erosion of beaches, damage to coastal habitats, and damage to foundations of infrastructure such as roads, railroads, bridges, buildings, and pipelines. Such damage can cost a fortune to repair or rebuild. Storm surge can also be deadly. Countless human lives have been lost to drowning in storm surges. But it's not only people who pay the price for hurricanes.

Hurricanes have a tremendous impact on our planet's wildlife. Animals suffer from the strong winds generated by hurricanes and can be displaced, injured, and killed or impacted indirectly through changes in habitat and food availability. Storms and dangerous waves can also move large amounts of sand and thus reshape coastal and **estuary** habitats, displacing or destroying wildlife. The impact on endangered species is even worse. For example, the Puerto Rican parrot—the only parrot remaining in the United States—saw its population reduced from 47 to 23 individuals after hurricane Hugo passed through the species' range in September 1989. Hugo ripped through Guadeloupe, Saint Croix, St. Thomas, Puerto Rico, and the southeast of the United States, lasting two weeks.

Saltwater intrusion from the storm surge travels far inland, contaminating freshwater lakes by changing the **salinity**, thus subjecting plants and animals to unsuitable living conditions. Turbidity from sediment and pollution is also a result due to back and forth water movement and runoff that impacts entire coastal habitats. Nesting sites for birds are also decreasing after a hurricane because of tree losses caused by the violent winds. Inland animals aren't the only wildlife affected by hurricanes. The ocean's ecosystem is also impacted, as sea animals are disturbed by the wind and waves that damage coral and other structures on reefs.

Figure 33 Puerto Rican parrot

Hurricanes kill millions of fish, both directly through the massive waves they create and indirectly, by rapidly altering the chemical balance and temperature of the water. Marine mammals such as whales, dolphins, and manatees may be injured or killed—crushed or beached by massive waves. They may also be swept from the calm, coastal waters that they prefer out into the open ocean, where they may become so disoriented and exhausted that they die before reaching the coast. For example, a pelican was blown all the way to Nova Scotia by Hurricane Earl in 2010, and two manatees were carried to a backyard pond after Irma struck Florida in 2017.

Changing habitats may also lead to shifting ecosystems and a disruption of the food web. Strong winds and flooding can destroy trees and other vegetation, leading to decreased food for surviving wildlife. Seagrasses are highly vulnerable to the increased flow of sediment caused by hurricane runoff. The grasses can be buried, and sunlight can be blocked from reaching the ocean floor, thus impacting the growth of new seagrass. Damage to sea beds can lead to numerous negative effects on marine creatures, including sea turtles, manatees, sea birds, **mollusks**, and other invertebrates. Members of all parts of the food web are affected by changes in food availability.

Much of the wildlife affected by hurricanes will eventually re-establish their areas, but it might not always be the case for exotics and invasive species, which may spread elsewhere and actually cause further damage to an ecosystem. Hurricanes are also linked to invasive species. Many invasive creatures may not seem very threatening, but these invaders, large and small, can have devastating effects on native wildlife.

Invasive Species[7]

An invasive species can be any kind of living organism—such as a fish, animal, insect, plant, fungus, bacteria, or even an organism's seeds or eggs—that is not native to an ecosystem and causes harm by competing with native species for the resources available in the area. The National Wildlife Federation explains that, "invasive species are among the leading threats to native wildlife. Approximately 42 percent of threatened or endangered species are at risk due to invasive species."

Scientists have recorded 1,207 different non-native aquatic species that were carried to new habitats by flooding that occurred during the 2017 hurricane season. Some have caused no problems in their new homes, but some, such as the giant apple snail—a native of Amazonia—has been outcompeting native species due to its voracious appetite for aquatic plants in the freshwater wetlands and waterways of peninsular Florida.[7] Once established, invasive species are very difficult to eradicate.

Human beings are the actors of a changing world. We can continue to contribute to climate change, or we can work to create a planet in which hurricanes are less frequent, climate change is slowing down via a reduction in greenhouse gases, and animals can be successful in their native habitats. All this can start with just a few actions taken by all of us, starting with you and me. Ride-sharing is something we can all try to do more often. Riding in a car alone isn't good for the environment because it pollutes a lot. Full cars have lower energy consumption per person and thus cause fewer CO2 emissions than a car with just one passenger.

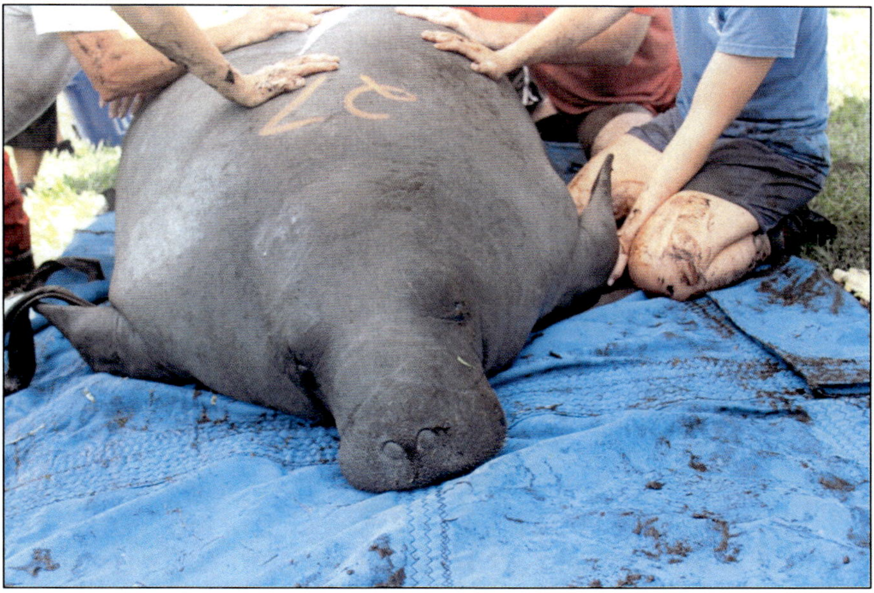

Figure 34 Stranded manatee rescued by Florida Fish and Wildlife after Hurricane Irma struck in 2017

We can also consider alternatives to cars. For example, we can try to use a bike or to walk for short-distance trips, and we can try to ride public transportation more frequently. Another thing we can try is using the sun to dry our clothes, thus reducing the amount of times we use gas or electric dryers. Buying from local producers is also one way to reduce emissions, since the food isn't travelling by plane or boat to land on your plate—it may come from only a few miles away. The provision and heating of water requires a lot of energy. For this reason, taking brief showers is more environmentally friendly than filling a bathtub. If everyone makes small efforts like this, at the end of the day, we will all see changes.

In addressing the issue of hurricanes specifically, one way to reduce the impact of hurricanes on coastlines is with mangroves, which "are tropical trees and shrubs that rest on the edge of a body of water, often where rivers meet oceans or seas."[1] With their thickly tangled roots, healthy mangrove forests located in both freshwater and saltwater areas have the ability to moderate the impact of flooding caused by hurricanes and tropical storms. They are very important resources, especially in coastal communities, where investing in mangroves could save millions of dollars in property and infrastructure damage.

In addition, a new idea known as the Salter Sink has been patented by British marine engineer Stephen Salter and billionaire philanthropists Bill Gates and Nathan Myhrvold. The plan for the Salter Sink "is to float thousands of tire-like rings in the tropical Atlantic, connected to giant tubes that suck warm surface waters down into deeper water, to be replaced by cold water from below."[9] This technique would prevent the water from reaching the required sea-surface temperature of 80 °F (26.7 °C) and thus prevent the formation of a hurricane. Another idea comes from engineer Mark Jacobson of California's Stanford University. Jacobson and his team believe that offshore wind farms could protect coastlines from approaching hurricanes. After going through recent hurricanes (Sandy in 2012, Isaac in 2012, and Katrina in 2015), they estimated that a huge wind **turbine** array off the coast of New Orleans could have reduced the peak wind speed of hurricane Katrina by an astonishing 90 miles (145 km) per hour and the accompanying storm surge by close to 80 percent.

Hurricanes are devastating natural phenomena. Climate change has led to a domino effect on the power and frequency of hurricanes, rising sea levels, and storm surges that impact both humans and wildlife. But even as natural disasters are becoming a greater threat to our planet, the solutions are right in front of us. If everyone is willing to contribute time and effort to understand the problem of climate change and take steps to address it, things will change and eventually move in the right direction for our planet. We can't be selfish regarding the situation of climate change and say, "I won't suffer from it because I won't be alive when the real consequences affect the Earth." We have to think about future generations, our grandchildren who will see the results of the damage we have allowed to happen. We are already fortunate enough to have people putting money, time, and effort into finding ways to combat the threat of hurricanes and other impending disasters. Let's all join the fight against climate change.

Figure 35 Highway destruction caused by Hurricane Katrina in 2005

1 Berezin, Jodie, et al. "Using Mangroves to Mitigate Hurricane Damage to the Southern U.S. Coast." *Debating Science*, 24 April 2018, blogs.umass.edu/natsci397a-eross/using-mangroves-to-mitigate-hurricane-damage-to-the-southern-us-coast.

2 "Climate Change: Meaning, Definition, Causes, Examples and Consequences." *YouMatter,* 27 April 2020, youmatter.world/en/definition/climate-change-meaning-definition-causes-and-consequences.

3 "Climate Kids." *NASA*, 18 June 2020, climatekids.nasa.gov/greenhouse-cards/#:~:text=Greenhouse%20gases%20are%20gases%20in,Carbon%20dioxide.

4 "Global Warming and Hurricanes." *Geophysical Fluid Dynamics Laboratory*, NOAA, 5 February 2020, www.gfdl.noaa.gov/global-warming-and-hurricanes.

5 "How Do Hurricanes Form?" *NASA Science,* 4 December 2019, spaceplace.nasa.gov/hurricanes/en/.

6 "Hurricanes." *United States Search and Rescue Task Force*, www.ussartf.org/hurricanes.htm.

7 "Invasive Species." *National Wildlife Federation*, www.nwf.org/Educational-Resources/Wildlife-Guide/Threats-to-Wildlife/Invasive-Species.

8 "Living in Coastal Areas." *World Ocean Review*, 2010, worldoceanreview.com/en/wor-1/coasts/living-in-coastal-areas.

9 McGuire, Bill. "How to Stop a Hurricane." *Science Focus*, 27 April 2018, www.sciencefocus.com/planet-earth/how-to-stop-a-hurricane/.

10 Rahmstorf, Stefan. "Rising Hazard of Storm-Surge Flooding." *Proceedings of the National Academy of Sciences of the United States of America*, 7 November 2017, www.pnas.org/content/114/45/11806.

11 "What is Storm Surge." *National Ocean Service*, NOAA,13 November 2019, oceanservice.noaa.gov/facts/stormsurge-stormtide.html.

SPECIES AT RISK: GRAUER'S GORILLAS

Oluwatobi Oladapo

"I am gorilla. I am flowers, animals. I am nature. Man Koko love. Earth Koko love. But Man stupid. Stupid! Koko sad. Koko cry. Time hurry! Fix Earth! Help Earth! Hurry! Protect Earth. Nature see you."

— Koko the gorilla

Human activity has become a bad seed planted in the ecosystems of wildlife all around the world. **Deforestation**, overfishing, global warming, pollution, and war are just some of the ways that humans have negatively affected the Earth. The Grauer's gorilla is a subspecies of gorilla that is bearing the brunt of these effects because human activities in their habitat is wreaking havoc on their population. This essay will break down the factors that have led these gorillas down their current path of a dying population and analyze the major roles played by humanity to speed up this process through selfish actions such as war, deforestation, and poaching. Many organizations are already aware of this problem and have started to act in order to offer hope for this species, but a reversal of the declining trend of their population lies in the hands of humanity and depends on quick action. As our closest genetic relative—and the reason for its impending demise—we owe this species a chance at a comeback.

In the heart of Africa lays the second largest rainforest after the Amazon. The Congo Basin Rainforest spans roughly 1.4 million square miles (3.6 million sq km) of equatorial Africa. This massive ecosystem is shared by six countries: Cameroon, Central African Republic, Gabon, Equatorial Guinea, Republic of Congo, and Democratic Republic of the Congo (DRC), with the DRC holding roughly 60 percent of the rainforest.

Contained in the rainforest is the second-largest river in the world, the Congo River. This rich ecosystem is covered in dense forest and swamp that is home to more than 600 tree species and more than 10,000 animal species, making its **biodiversity** one of the most complex in the world. Gorillas, elephants, chimpanzees, okapi, leopards, hippopotamuses, and lions are just some of the many animals that can be found in the area. In fact, the forest itself provides shelter for a variety of gorillas. The two distinct species of gorilla—western and eastern—are divided into four subspecies. The Cross River gorilla (*Gorilla gorilla diehli*) and the western lowland gorilla (*Gorilla gorilla gorilla*) are subspecies of the western gorilla. The eastern lowland gorilla (*Gorilla beringei graueri*) and the mountain gorilla (*Gorilla beringei beringei*) are subspecies of the eastern gorilla.

The eastern lowland gorilla is also known as Grauer's Gorilla, named after Austrian scientist Rudolf Grauer, who first identified it as a subspecies. Males average 5 to 6 feet (1.5–1.8 m) tall and weigh up to 500 pounds (227 kg). Females are generally smaller than males; they average just under 5 feet (1.5 m) tall and weigh 155 to 200 pounds (70–90 kg). These gorillas are **endemic** to the Albertine Rift in the eastern part of Democratic Republic of the Congo. They live both in lowland and mountain tropical broadleaf forests, where they can eat up to 40 pounds (18 kg) of food in a day. They are mostly **herbivorous**, eating stems, bamboo shoots, fruits, and leaves. On rare occasions, they may feed on insects. Grauer's gorillas live together in family groups called troops. A dominant male, called a silverback, assumes the position as the leader of the family, which consists of as few as 2 or as many as 40 members.

Terms to Know

Mature males from 8 to 11 years old are called **blackbacks**. When males reach about 12 years of age, white hair covers their backs and shoulders, and they develop sharp canine teeth. These gorillas are known as **silverbacks**.

These **nomadic** apes spend most of the day foraging for food or sleeping in temporary nests that they make in trees or on the ground. They are typically calm and gentle; nevertheless, a silverback will not hesitate to defend his family against threats or opponents. His duty is to protect the females and their offspring, lead them to feeding spaces in the forest, and mate with females to produce more offspring. Gorilla populations grow slowly. Females begin mating around age 10, typically giving birth once every 3 to 4 years. They have only 2 to 6 offspring over a lifespan of about 40 years.

Figure 36 Rainforests are some of the most biodiverse ecosystems on Earth.

Grauer's gorillas are important to the ecosystem in which they live and can even be representative of their environment's stability. They play a major role in the food web, as they are driving forces of the cycle of life in these forests. As herbivorous, nomadic animals, they are able to support the growth of forests by dispersing seeds in their dung. They eat a more diverse selection of fruits because they are able to swallow seeds that are too big for most other animals. This ability helps maintain diversity and balance throughout the ecosystem because they can affect the new growth of trees, and are thus directly connected to the sustainability of other species in the ecosystem. The loss of Grauer's gorillas would undoubtedly alter nature in the Congo Basin.

These animals are not only useful to their native habitat, but they are also important to human science. They share 98.5 percent of our DNA, making them our closest biological relatives. This is beneficial to us because scientists rely on them in biological, medical, and evolutionary research. For example, in 2004 the researchers of the Wildlife Conservation Society used information from gorillas that were exposed to the Ebola virus to study the distribution of the virus. Clearly, if we lose Grauer's gorillas, we would be losing a critical member of our global community.

In the past two decades, the population of Grauer's gorilla has declined so much that the International Union for Conservation of Nature (IUCN) Red List now lists them as critically endangered. There were 16,900 Grauer's in the mid-1990s, but after recent surveys report that their population has since declined by 77 percent. In fact, the last known census estimated only 3,800 individuals left in the wild, and this number could be lower because it has been difficult for proper and accurate surveys to be taken.[10] This drastic decline in the Grauer's gorilla population has also led to their reduced and fragmented range across eastern DRC. They have been reported to now occupy only 13 percent of their historic range and are highly fragmented in these areas.[4] The Grauer's gorilla now survives in only a few DRC locations where conservationists and researchers have been working to protect them. These include Maiko National Park, Punia Gorilla Reserve, Kahuzi-Biega National Park, Itombwe Reserve, and a few more places. Even in these legally protected areas, the gorillas are not safe, and their future continues to be threatened.

Grauer's gorillas have long been under siege on all fronts by human activities. After the Rwandan Civil War and **genocide** against the Tutsi that occurred in 1994, thousands of refugees rushed into the eastern parts of Congo (where Grauer's gorillas live), sparking more conflict in DRC. Corruption and severe debt caused by the lack of proper leadership from the current leader did not allow DRC to develop economically or socially after it became independent of Belgian **colonialism**. By 1996, war broke out in DRC, and there were two civil wars that raged until the year 2003.

Figure 37 Gorillas look like fierce creatures, but they are actually quite calm.

The conflict prevailed mostly in the east around the mining regions of DRC where several rebel militia groups engaged in violence against each other for these resources. In the 2000s, there was a transition of government that maintained the corrupted patterns of the past leaders. The situation got so bad that by 2009, there were only 5 countries below the DRC in the human development index (a measurement of various achievements, including a long and healthy life, being knowledgeable, and having a decent standard of living).[3] DRC has always been a country plentiful in resources, so it was no surprise that other countries wanted a piece of the pie. In 2007, China got into business with DRC, setting up an agreement that would allow China to extract resources in exchange for capital for investment in infrastructure. This was a good plan to bring DRC out of debt and help it flourish economically, but corruption in the government and the armed militia combined with widespread poverty have stunted the growth of DRC.[3]

Some of DRC's many resources include gold, diamonds, copper, uranium, and coltan (a mineral used to make mobile phones). Once these valuable raw materials were discovered after the war, the armed rebel militia also invaded the forests, taking control of Kahuzi-Biega National Park, a key area that had offered Grauer's gorillas some measure of protection under the watchful eye of the Institut Congolais pour la Conservation de la Nature (ICCN), the Congolese Wildlife Authority.[8]

Figure 38 Mature male gorillas are called silverbacks.

Many extractive industries established their own mines causing human activity in these undisturbed areas to increase, and the government set up large mining sites all over the region to extract these resources and make money by exporting it to other countries. When resources are extracted from the ground, large patches of forest get cut down for mines, and even though the government makes money off these transactions, many people in the society are still living below the average socioeconomic status. For the survival of their families, people who live in these rural areas have no choice but to engage with these industries by becoming farmers and miners, causing more negative attention to be drawn to the habitats of the gorillas. Some of the problems that have befallen the wildlife in these forests center mainly on overexploitation that range from habitat loss caused by illegal artisanal mines controlled by armed militia and deforestation for farming and fuel to poaching and the illegal wildlife trade. These threats ultimately leave the Grauer's gorilla population fragmented across their range, which leads to a continual decline in their numbers.[11]

A major issue related to artisanal mining is the illegal hunting and killing of gorillas for **bushmeat**, which miners use to feed their families—often simply because of a lack of affordable alternatives. A report funded by the Arcus Foundation stated that "many of the miners, aware that bushmeat hunting has led to the decline of some mammal populations, would prefer other food sources, and many would like to find steady sources of income outside of forest exploitation."[5]

Due to the armed militia and army roaming the forests intent on making money off the wildlife, a secondary threat is poaching. Sonya Kahlenberg, U.S.-based executive director of the Gorilla Rehabilitation and Conservation Education (GRACE) Center, west of the Virunga National Park in the eastern part of the Congo Rainforest, explained that "it's not uncommon for armed groups to kill adult gorillas for food and take the babies to sell on the **illicit** market or to keep as trophies or pets." Kahlenberg further explained that reintroducing eastern gorillas to the wild is typically not done, the logic being that "once removed from their mothers, gorilla babies cannot return to the wild because most are not yet weaned and require supplemental nutrition." For example, when Muyisa, a four-year-old female gorilla was brought to the center, Muyisa adapted just fine to the small population of gorillas in the sanctuary, but she would never again know life in the wild.[5]

The deadliest threats are often those that cannot be seen—those that hide under the skin, leaving an organism incapacitated without physically observable evidence. This type of threat affecting the Grauer's gorilla is its rapidly reducing gene pool. A recent study done by Katerina Guschanski and her colleagues at Uppsala University provided evidence to explain this obscure phenomenon. After sequencing genomes from 59 museum specimens of older generations of mountain and Grauer's gorillas collected between 1910 and 1962, as well as genomes of modern mountain and Grauer's gorillas, Guschanski's team found that there was a notable discrepancy in their genomes. The two types of gorillas were very different genetically. Almost 10,000 years ago, the eastern gorillas split into the mountain gorilla and Grauer's gorilla subspecies, divided by the type of tropical forest in which they lived.[7]

After the split, the Grauer's gorilla population grew greatly; however, when the rapid population decline of the Grauer's gorilla started 20 years ago, it caused random genetic changes that bottlenecked the species' population. Guschanski's project showed that this speedy crash of the species population has left Grauer's gorillas at a risk of lower genetic diversity. The newer generation of Grauer's gorillas have lost their heterozygosity, meaning that they have become more genetically similar than their ancestors were, leading the average modern Grauer's gorilla to have lost 20 percent of its genetic diversity in only 4 to 5 generations.[2]

The implication of this revelation is that the Grauer's gorillas now have genomes with marks for **inbreeding**, and the rapid growth of the Grauer's gorilla population before its decline caused the subspecies to develop harmful genetic mutations that are now ubiquitous in the modern species, leaving the entire population vulnerable and forcing it into a position of requiring it to adapt quickly to changing environments with a less effective genetic toolkit.[7]

Figure 39 Grauer's gorillas are have no defense against genetic problems.

Mutations in DNA can be very harmful to a species because they may lead to serious problems such as weakened immune systems and reproductive deficiencies, leaving a species weak and unable to adapt to changes in its environment. "If you've only got 200 individuals left, or even a thousand individuals, one bad disease or one nasty episode where a lot of them get poached or killed could wipe them out," University of Cambridge geneticist Aylwyn Scally explained.[2] The existence of mutation and inbreeding within this subspecies of gorilla, along with all of the other threats it faces, can accelerate the pace of its already declining population. Despite this, scientists hold out hope that the Grauer's gorilla will eventually adapt to its low genetic diversity. Research on the genetic development of the species continues on several fronts.

With the possibility of weakened immune systems, disease also poses a threat to Grauer's gorillas. Research conducted by Murdoch University PhD student Alisa Kubala stresses the importance of understanding how pathogens move between humans and **primate** species. She explains that "humans and gorillas share 98.5 percent of their genetic makeup, which makes them susceptible to many of the same infectious diseases. However, while humans have traveled around the world and built up resistance to many diseases, gorillas remain isolated and immunologically naïve and are at greater risk of acquiring new infections." By collecting fecal and blood samples in the Kahuzi-Biega National Park and the Virunga National Park, Kubala studied the transmission of malaria, microfilaria, and HIV's ape equivalent Simian Immunodeficiency Virus (SIV) between humans and gorillas.[9]

Research is vital as the first step to take when considering the efforts for conservation of any species. It is necessary to find out how badly the species has been affected, and that will not be determined until information has been gathered concerning various aspects of the species, its population, and its habitat. Past initiatives to conduct ground studies and surveys of the Grauer's gorillas in DRC have faced difficulty, mostly as a result of the high level of insecurity in the habitats of the gorillas. One main problem to overcome is that the armed militia groups in these regions resort to violence when the territories they control are invaded. In 2016, two forest rangers, Oscar Byamungu Mianziro and Munganga Nzonga Jacques, were killed by the armed militia in the area while they were patrolling the park. Violence like this is just one of the various obstacles that must be tackled if researchers and rangers wish to conduct studies and patrol the forests—even in the areas such as Kahuzi-Biega National Park, which have been awarded legal protection. Scientists are limited in the areas they can locate for surveys and must often walk intensive routes with patrolling rangers to avoid danger. Due to the violence, these rangers risk their lives and often find themselves in armed confrontations with the rebel militia in the name of conservation.

Amid the hardship, some surveys have been successful in tracking existing groups of Grauer's gorillas in order to determine their current status. The scientists that embark on these expeditions are faced with the challenge of physically counting the animals because their reduced nomadic population makes it hard for scientists to track them down and give accurate estimates. Prior to 2011, very little was known about the status of Grauer's gorillas; the Wildlife Conservation Society had not done a survey since the early 1990s. But in 2011, the Jane Goodall Institute, with the help of 11 other organizations including the Wildlife Conservation Society, Fauna & Flora International, and the Institut Congolais pour la Conservation de la Nature, took the initiative to develop a participatory Conservation Action Plan for Great Apes in Eastern DRC.

In previous years, the IUCN had classified Grauer's gorillas as endangered, but the survey was proposed as an effort to increase the urgency centered around these gorillas. These organizations wanted the IUCN to elevate the status of Grauer's gorillas from endangered to critically endangered so that the efforts to save them could increase. Preliminary research from some of the first surveys on Grauer's gorillas in the second half of the 20th century confirmed that the species' population was declining. The later surveys in the 1990s centered mainly around the large populations of the Grauer's in Kahuzi-Biega National Park, Maiko National Parks, and Itombwe National Reserves, which estimated that the remaining population was 16,900. The 2011 report compiled and summarized current information based on the surveys taken and explained the causes of the decline of the population. As a result of the survey, in 2012, the scientists estimated that the population of Grauer's gorillas in all regions of their existence was between 3,000 and 7,000. Over the next five years, the more field surveys were conducted with the help of local communities. They used technology to track the gorillas and collect data on threats posed to the animals, such as poaching and habitat loss. In 2016, the final report was published; it identified that the estimated total population of Grauer's gorillas was 3,800. The report led the IUCN to elevate the status of the gorillas to critically endangered and brought to light the Grauer's gorilla's dire situation.[10]

Fauna & Flora International researcher and co-author in the study Stuart Nixon stated, "As one of our closest living relatives, we have a duty to protect this gorilla from extinction. Unless greater investments and effort is made, we face the very real threat that this incredible primate will disappear from many parts of its range in the next five years. It's vital that we act fast."[11] Nixon wasn't the only one who had something to say about the researchers' findings. The lead author of the Wildlife Conservation Society's research, Andrew Plumptre, urged the government to "quickly establish Reserve des Gorilles de Punia and the Itombwe Reserve, and reinforce Kahuzi-Biega National Park efforts, which have community support, and to establish strong coordination between ICCN and the DRC military to tackle armed militias that control mining camps in Grauer's gorilla heartland."[11]

Since the report was published, efforts for the conservation of Grauer's gorillas have begun to intensify. One such organization to take up the challenge is World Wildlife Fund (WWF). Founded nearly 60 years ago, WWF has been involved actively in the conservation of Grauer's gorillas for a long time. In 2007, WWF and some partner organizations succeeded in getting nine African countries, including DRC, to sign a legally binding agreement to fight poaching.[10] WWF also got 10 countries to collaborate in the fight against threats to gorillas under the Gorilla Agreement, signed in June 2008.[12] The two main objectives are to monitor the gorillas and to expand protected areas. After the civil wars, they have since trained and funded many Congolese Wildlife Authority (ICCN) staff to carry out more surveys and patrol the Kahuzi-Biega National Park to protect it from poachers and illegal mining. As a result, the ICCN staff now control over 70 percent of the Kahuzi-Biega National Park compared to 10 percent in 2003.[1] Such efforts also emphasize the importance of protected areas as a necessity for conservation and the fight against habitat loss. To achieve this goal, they are contributing to the local communities by providing food, water, and education to train the people in methods of protecting their wildlife, and they are pressuring the government to identify and establish more protected areas that will be kept safe from poachers and violent groups.[4] The Tshivanga highlands in the Kahuzi-Biega National Park is a testimony to their success. A small population of 130 gorillas in 2000 grew to 213 by 2015, showing a 64 percent increase.[8]

Figure 40 Young Grauer's gorillas are particularly vulnerable to human interference.

A number of organizations around the world are actively joining forces to reverse the population decline of Grauer's gorillas. The main objectives of conservation efforts are to legally define boundaries of protected areas, stop illegal mining, provide alternative sources for livelihood of the community people, disarm the rebel militia in the area, raise more staff and field workers that can protect the habitats of the gorillas, and lobby for the government to produce legislation that terminates illegal mining, poaching, and violence in the area. Specific efforts are focused on the regions that contain the main populations of Grauer's gorillas: Usala Forest; Kahuzi-Biega National Park; and Punia Gorilla, Itombwe, and Tayna reserves.[11]

Also, the Dian Fossey Gorilla Fund collaborates with local communities to protect Kahuzi-Biega National Park as well as the forests in the Nkuba Conservation Area (named for a nearby village), where almost 200 Grauer's gorillas live.[6]

Name to Know

Dian Fossey (1932-1985) was an American **zoologist** who became the world's leading authority on the mountain gorilla. She conducted research and led conservation efforts in Rwanda. The Dian Fossey Gorilla Fund is the largest and longest-running organization fully dedicated to gorilla conservation.

Funds are vital for all these organizations to continue their work. The United Nations Education, Scientific and Cultural Organization (UNESCO) works with the World Heritage Committee to approve requests for funds to be sent to these projects. While UNESCO has sent $141,970 USD to the Kahuzi-Biega National Park between 1980 to 2018, this amount is just not sufficient to fully change the situation in DRC.[13] Each project involves a variety of logistics, all of which take massive funding.

More can be done to save Grauer's gorillas and the Congo Rainforest that they call home. Humans caused the challenges that the gorillas face, so we owe it to these animals to keep them alive. This can only be done if the government in DRC can unite with the organizations around the world that are trying to help. Every entity involved must be on the same page to ensure that proper and efficient action is taken in the Congo Rainforest. The organizations that have stakes in Grauer's gorilla conservation efforts suggest many plans to rescue this dying species, but I believe that to be truly successful, they should all be summarized and presented as a whole. Two hands are better than one. All of these organizations and the government joining efforts together to initiate one plan could successfully propel conservation and reverse the decline of Grauer's gorilla numbers. The IUCN does a good job of presenting a plan that covers all fronts of the issues that the gorillas face today.

The first thing to be done should be to increase public awareness in DRC and all around the world. This will help with raising funds for the government of DRC to put policies in place that can actually be implemented. These funds will have to be invested into multiple aspects such as educating the people on the situation at hand, building the socioeconomic status of people who are involved in the exploitation of the forest resources, providing alternatives for those involved in extractive industries, and increasing the security of the forest. The security is a major part of the plan because it ensures the disarming of militia in the area and terminates poaching operations, ultimately keeping sanctuaries and protected areas safe. Land must be demarcated and well planned so that the government can keep human activities from further destroying the wildlife in the Congo Rainforest. After all this has been done, there must then be proper patrolling and monitoring of the Grauer's gorillas to make sure that their population is kept safe and allowed to grow.

The dwindling population of the Grauer's gorilla, a valuable member of the great apes and our closest genetic relative, is facing the dire repercussions of unrestrained human activity. Grauer's gorillas are intertwined with their habitats and the wildlife around them; as a result, the health of their population greatly reflects the ecosystems where they live. Since human activity such as deforestation, mining, and poaching have greatly contributed to their decline, it rests upon humanity to reverse this devastating trend. Many organizations have taken up the torch and begun to address the situation, but more action needs to be taken to produce adequate results. While governments and conservation organizations can implement plans that will raise awareness, funds, and security for the gorillas, as a member of humanity, you are tasked with providing for the planet just as it provides for you. Even if you can't donate funds or time to conservation organizations, you can and should choose actions on a daily basis that demonstrate respect for other living things and, ultimately, the Earth as a whole.

Figure 41 Grauer's gorillas need our help.

1 "African Great Apes Solutions." *WWF*, WWF–World Wide Fund for Nature, 2020, wwf.panda.org/knowledge_hub/endangered_species/great_apes/apes_programme.

2 Cannon, John C. "Rapid Population Drop Weakened the Grauer's Gorilla Gene Pool." *Mongabay Environmental News*, 10 January 2019, news.mongabay.com/2018/11/mountain-gorillas-recovery-prompts-downlisting-to-endangered-status.

3 Crash Course. "Congo and Africa's World War: Crash World History 221." *YouTube*, 16 January 2015, youtu.be/4uArRzwKHvE.

4 "Eastern Lowland Gorilla." *WWF*, World Wildlife Fund, 2020, www.worldwildlife.org/species/eastern-lowland-gorilla.

5 "Endangered Eastern Lowland Gorilla Orphan Finds New Home." *Arcus Foundation*, 25 July 2017, www.arcusfoundation.org/stories-of-impact/great-apes/endangered-eastern-lowland-gorilla-orphan-finds-new-home.

6 "Grauer's Gorillas." *Gorillafund*, Dian Fossey Gorilla Fund International, 10 April 2020, gorillafund.org/grauers-gorillas.

7 Lewis, Ricki. "Museum Genomes Explain the Plight of the Grauer's Gorilla." *DNA Science Blog*, Plos.org, 10 January 2019, blogs.plos.org/dnascience/2019/01/10/museum-genomes-explain-the-plight-of-the-grauers-gorilla.

8 "Measures." *WCS*, Wildlife Conservation Society, 2020, measures.wcs.org/Metric-Details/m/24?_ga=2.76130453.1403317077.1587834462-917324734.1583902331.

9 Murdoch University. "Investing the Health Threats to Endangered Eastern Gorillas." *Phys.org*, 1 October 2014, phys.org/news/2014-10-health-threats-endangered-eastern-gorillas.html.

10 Plumptre, Andrew J., et al. "Status of Grauer's Gorilla and Chimpanzees in Eastern Democratic Republic of Congo." *Arcus Foundation*, Wildlife Conservation Society, April 2016, www.arcusfoundation.org/wp-content/uploads/2016/04/FFWCS.pdf.

11 Rakowski, Sarah. "New Study Documents Shocking Collapse of Gorilla Subspecies During 20 Years of Civil Unrest." *Fauna & Flora International*, 4 April 2016, www.fauna-flora.org/news/new-study-documents-shocking-collapse-of-gorilla-subspecies-during-20-years-of-civil-unrest.

12 "Solution." *WWF*, WWF–World Wide Fund for Nature, 2020, wwf.panda.org/knowledge_hub/endangered_species/great_apes/gorillas/save_solutions_gorillas.

13 UNESCO World Heritage Centre. "UNESCO World Heritage Centre – State of Conservation (SOC 2019) Kahuzi-Biega National Park (Democratic Republic of the Congo)." *UNESCO*, UNESCO World Heritage Centre, 2019, whc.unesco.org/en/soc/3844.

APPALACHIAN RED SPRUCE FORESTS: DESTRUCTIVE PAST TO OMINOUS FUTURE

Adam L. Osborne

"A nation that destroys its soils destroys itself. Forests are the lungs of our land, purifying the air and giving fresh strength to our people."

— Franklin D. Roosevelt

Red spruce (*Picea rubens*) once dominated the high-elevation ecosystems of the central and southern Appalachian Mountains. Prior to the nineteenth century, vast stands of pure red spruce occurred in great abundance throughout this area. These forests contained ample wildlife and ecological diversity. Over the past century, due to a combination of factors, that range has dwindled to a small fraction of the original size. Along with the reduction of forest, numerous animal species have also become threatened or endangered. Animals that have been directly affected by the reduction of red spruce include the pigmy **salamander** (*Desmognathus wrighti*) and the northern flying squirrel (*Glaucomys sabrinus*). The future of these areas, plagued by acid deposition and climate change, is unknown. This "riches to rags" story, which begins centuries ago, follows a remarkable path that coincides directly with the expansion and advancement of the human race. This passageway through time explores uncharted virgin forests, uncontrolled logging, destructive fires, salvation efforts, and an ominous future. With the combined effects of previous destruction and future problems, is there hope for the red spruce?

Ancient Wonders

Before understanding where the red spruce is going, one must first understand the red spruce itself. Red spruce is an evergreen tree, meaning that the tree retains needles and stays green year-round. Red spruce is part of the Pinaceae family of plants. Red spruce is found in high elevation climates that are over 3,500 feet (1,067 m) in elevation. Red spruce favors the cool, moist climate conditions that occur in these areas. In the central and southern Appalachian region, these climates, ones over 3,500 feet (1,067 m), are hard to come by. Harold S. Adams et al. states in a United States Forest Service article, "[r]ed spruce has a restricted distribution in the mountains of central and southwestern Virginia and eastern central West Virginia. Presumably, distribution is limited because few areas in this region of the Appalachians reach the elevations necessary to provide the cool, moist conditions that this species requires to grow and reproduce."[1] The red spruce already has a disadvantage in these areas from the start, and as we'll see, it doesn't get much easier from here.

The red spruce, and gymnosperms alike, can be traced back millions of years and are some of the oldest living organisms on Earth. Gymnosperm is a word that means "naked seed." This description comes from the fact that their seeds are exposed on woody "bracts" or cones. Red spruce trees will grow, provided with favorable conditions, to a height of 100 feet (30.4 m) or greater. These trees have the ability to create a very dense forest. This is due to the trees having an excurrent growth form. An excurrent growth form exhibits apical dominance; in other words, the upper stem outgrows the lateral branches. This results in a cone-shaped tree, one that is much taller than it is wide. This allows the trees to grow in close proximity to one another, thus creating the dense stand. This closeness of growth also dramatically reduces the amount of sunlight that can reach the forest floor. This creates a dusk-like habitat, even in the middle of the day.

An Ideal Abode

Red spruce forests provide a home for a vast array of animal life and **ecological** diversity. If we look at where these forests are located in proximity to their surrounding locations, we see a pattern. These forests serve as isolated "islands" that contain their own unique species and wonders. These forests are home to large animals such as white-tailed deer (*Odocoileus virginianus*) and American black bear (*Ursus americanus*), while also containing countless small animal species such as the northern flying squirrel, pigmy salamander, and Cheat Mountain salamander (*Plethodon nettingi*). These animals have called red spruce forests home for generations. The cool, moist climate conditions that occur here are a prime environment for these species. They have evolved to survive exceptionally well in these areas.

The northern flying squirrel is found throughout the eastern United States in areas other than the central and southern Appalachian Mountains. When looking on a regional basis, in these parts of the Appalachians, it is dependent upon the climate that surrounds the red spruce. It is unlikely that the species as a whole will go extinct; however, the small populations of subspecies are more vulnerable and have greater potential to be harmed.

Figure 42 Northern flying squirrel

Eldad Malamuth and Michael Mulheisen state in an *Animal Diversity Web* article "the subspecies *Glaucomys sabrinus coloratus* and *G. s. fuscus* are threatened populations in the Appalachians." The logging and subsequent destruction of suitable habitat directly affected the flying squirrel. Malamuth and Mulheisen go on to explain "between the 1880's and the 1920's, 500,000 acres [202,343 ha] of forest supporting the two subspecies were reduced by timbering to 200 acres [81 ha]. Conservationists are concerned that further habitat destruction, fragmentation, and pollution will eliminate the small and vulnerable islands of high elevation habitats."[3] The northern flying squirrel is a species that, while perhaps not a concern in certain areas of the country, should be of the utmost importance in forests of the central and southern Appalachian Mountains.

Another type of animal that is affected by the health and expanse of the red spruce is the salamander, specifically two species of salamander. The pigmy salamander and the Cheat Mountain salamander are two species that are dependent upon the red spruce forests. The pigmy salamander is endemic to high-elevation red spruce forests in the mountains of central and southern Appalachia. The loss of suitable habitat has had a direct effect on the pigmy salamander. A *Tennessee Wildlife Resource Agency* article states "populations are in jeopardy due to loss of habitat from logging activities, increased recreational development, and stream pollution."[6] This helps to explain that while timber loss may have sparked the decline, other factors have also assisted the degradation of the habitat for the pigmy salamander.

The Cheat Mountain salamander is also directly tied to the red spruce forests of the Appalachian Mountains, and more specifically the forests of West Virginia. Thomas K. Pauley states in a *Northeastern Naturalist* article "the original *Picea rubens* (Red Spruce) forest in West Virginia covered approximately 1.5 million acres [607,028 ha], most of which was eliminated between 1870 and 1920 by clear-cutting and conflagrations. The total range of *Plethodon nettingi* (Cheat Mountain Salamander) was confined within this Red Spruce forest."[5] This salamander is special in that it is currently found solely on Cheat Mountain and nearby surrounding mountains in West Virginia. The widespread logging that occurred around the turn of the twentieth century and the effects that came with the logging would devastate the salamander. The widespread fires that were greatly fueled by the clear-cuts further reduced the suitable habitat for the salamanders. Thomas K. Pauley goes on to state "in the once abundant Red Spruce forests in eastern West Virginia, leaf litter decay was slow due to the acidic nature of spruce needles, which resulted in the formation of a thick layer of partially decayed needles and twigs several feet deep."[5] This large amount of "duff" on the forest floor was left following logging. This newly exposed material was then allowed to dry in full sunlight. This created a tinder box waiting for the match to be lit. Fires would destroy the remaining material on the forest floor, upon which the salamanders depended to survive.

Figure 43 The Dolly Sods Wilderness in the Allegheny Mountains of eastern West Virginia is included in the Cheat Mountain salamander's range.

Figure 44 Pigmy salamander

Figure 45 Cheat Mountain salamander

Losing Ground

For thousands of years, the red spruce forests of the central and southern Appalachian Mountains went untouched by the hand of man. This was due, in part, to being located in a very rugged landscape. In these mountainous areas, it is not uncommon for the elevation to reach 5,000 feet (1,524 m) or greater. These mountains are also characterized by having great variance in elevation as well as large rock outcroppings and many streams that form a sort of web throughout the mountain landscape. Prior to the arrival of the first settlers in the eighteenth century, these forests were primarily virgin. The first foray in these lands were centered around rivers. Early settlements were concentrated around waterways. This is due to this area of land having the most negotiable terrain and having a means of transporting goods. The red spruce trees found along the rivers were the first to feel the result of human expansion as they were cut down to provide open land for livestock and settlement.

The settlers of the early days were limited by the equipment available. Small hand tools proved no match for the gigantic, old growth red spruce. Unable to directly cut down the trees, the settlers turned to another method to achieve their goal: the use of girdling. Girdling a tree is the act of cutting a shallow ring around the circumference of the trunk. This severs the cambium layer and xylem cells of the tree. The cambium of a tree is the actively growing tissues that allow the tree to increase in size. The xylem cells of the tree are little interconnected highways that transport water from the roots to the actively growing canopy. If these highways are cut off, then the tree has no means to grow. The backbone of tree growth is **photosynthesis**, which occurs in the leaves. Without water, photosynthesis cannot occur, and without photosynthesis, a tree cannot reproduce and grow. Once the girdling had accomplished its goal, the fate of the tree would become clear. The same restrictions that prohibited harvesting of the large trees would prevent the utilization of the wood. Once the trees had been killed, they were usually burned on the forest floor. Thousands of years' worth of wood was allowed to smolder in piles on the same land that it had grown.

Although destructive and wasteful, the girdling of red spruce was not a defining factor that led to its decline. It was not until the late nineteenth century that the real players became known. It was around this time, and throughout the turn of the twentieth century, that vast improvements were made in the field of timber harvesting. Inventions such as the band saw and circular sawmill allowed for expedited wood processing. The steam locomotive provided for eased transport of timber from the forest to the mill. These advancements also ushered in a new era for the red spruce. Logging companies now had the ability to harvest vast stands of timber.

By this time in history, the areas surrounding these mountains had become void of merchantable timber due to the boom in human expansion. The red spruce forests were the last means of untouched timber in the area. Logging companies, needing to supply their mills, turned to these old-growth forests. Gregory Nowacki et al. states in a United States Forest Service article "the appearance of modern technologies coupled with the existence of vast red spruce resources unleashed an unprecedented level of logging and environmental destruction along the Appalachian Chain."[4] This logging boom, and factors thereafter, would prove too much for the red spruce.

Timber companies of the day were interested solely in making money. Unlike timber harvesting of the modern day, there were no environmental restrictions that the loggers had to abide by. This allowed the loggers to do whatever they wanted. Clearcutting was the preferred harvest method of the area and time. It allowed for the greatest revenue for the loggers. Clearcutting is harvesting an area of forest down to a very small diameter limit or no limit at all. This essentially "clears" the land of trees. Clearcutting, in and of itself, is not all bad. When utilized correctly, with the correct tree species, clearcutting can achieve excellent regeneration, which is exactly what happened. To expand on the destruction facilitated by the logging, post-harvest fire would add to the decline of the red spruce.

Figure 46 Red spruce seeds

Nowacki further states "multiple fires were particularly devastating to red spruce because it is a non-sprouting species. These wildfires greatly reduced or eliminated red spruce on many landscapes, consuming its fire-susceptible seedlings, saplings, and remaining seed trees while rendering site conditions inadequate for regeneration. Under these circumstances, hardwoods benefitted vastly at red spruce's expense."[4] Hardwood tree species are much better suited to a fire-affected ecosystem in this area. This allowed the red spruce to be replaced by hardwood tree species. The fires allowed for excellent regeneration, although the majority of that regeneration was in the form of hardwoods. The mountains that were once covered in dense red spruce forests now are diverse hardwood forest communities.

If we look at a somewhat more recent history, we see that climate change and acid deposition have greatly impacted the red spruce. Acid deposition is a particularly interesting area of concern. Throughout the twentieth century, the burning of coal in coal-fired power plants created emissions that would combine to form acid rain. A majority of these plants were located upwind of large red spruce forests. This allowed the acid to be carried directly to the red spruce forests. Acid deposition can have very harmful effects on many plant species. Matthew B. Rosenberg and David J. Butcher explain in an *Instrumentation Science and Technology* article "a decline of red spruce forests in the southern Appalachian Mountains caused by atmospheric acid deposition has been reported since the 1970s. Acid deposition leaches essential nutrients (calcium and magnesium) out of the soil and increases the availability of toxic metals (aluminum) to plants."[7] These effects are coupled with the fact that they can be dispersed over a wide area. Climate change and warming climate conditions have had an effect on the red spruce as well, although this is more of a concern for the future than it has been for the past.

Climate change and warming conditions are a very bad combination for the red spruce. Red spruce needs cool, moist climate conditions to grow well. These optimum growing areas have been slowly disappearing, and by all indications, they will continue to do so. Over the history of the red spruce, they have slowly "retreated" to find these coveted growing conditions. We now find the red spruce with nowhere left to retreat. These trees have reached the point where they have no higher elevations to "climb" to. They are, in a word, stuck. Having retreated to the highest peaks available, they now wait to learn their fate. As the warming temperature slowly creeps up the mountains, ever-increasing amounts of suitable habitat is gobbled up. Potentially, the red spruce may run out of suitable habitat in these areas.

Every Little Bit Helps

The red spruce has been the focus of countless organizations over the past. One of the biggest "wins" for the red spruce was the creation of national forests. These areas saved small pockets of virgin red spruce from being destroyed. These areas also created sanctuaries where restoration efforts could be accomplished. Efforts have been made to restore red spruce in the form of planting projects. One such program is the Central Appalachian Spruce Restoration Initiative (CASRI). A *U.S. Forest Service* article states "to date CASRI has raised over $160,000 in grant money for on-the-ground projects, [and] planted over 85,000 native trees on both public and private land."[2] This is just one example of successful restoration work regarding the red spruce.

Red spruce has made a sort of comeback over the last century or so. Red spruce currently occupies the largest area since the start of logging. This is due in part to restoration projects and increased public knowledge of the forests, but also to the fact that nature is an excellent mediator. When allowed to function without human intervention, natural succession can work wonders. Slowly, over time, red spruce has taken over areas that were once lost. If given enough time, without exterior factors such as climate change, many tree species will naturally extend their range. This is evident with the red spruce.

If we look at the future of the red spruce, we see that climate change is the one central issue. Even with all of the restoration efforts, if there is limited suitable growing conditions, only so much can be done. Climate change is an issue that cannot be solved overnight. This is an area that will require years of debate and progress before noticeable improvements can be seen. The issue of the uncontrolled logging may have found a savior in the form of restorative tree plantings, but climate change is a different story.

A Long Way to Go

The vast majority of this essay has told the story of destruction and a bleak future. While this is, indeed, the current situation, it is not the entire situation. The goal of this research was to answer whether or not there is hope for the red spruce. I believe that there is hope. While that hope may be clouded by destruction, climate change, and unknown future, there is hope. We see that humans are very resourceful. The same species that is responsible for nearly wiping out the red spruce is now the its greatest ally. While the road ahead is rocky, at least there is a road.

---

1 Adams, Harold S., et al. "The Isolated Red Spruce Communities of Virginia and West Virginia." *Proceedings from the Conference on Ecology and Management of High-Elevation Forests in the Central and Southern Appalachian Mountains*, U.S. Forest Service, United States Department of Agriculture, 2010, www.nrs.fs.fed.us/pubs/gtr/gtr-p-64papers/01-adams-p-64.pdf.

2 "Central Appalachian Spruce Restoration Initiative (CASRI)" *U.S. Forest Service, United States Department of Agriculture*, www.fs.usda.gov/detail/mnf/workingtogether/partnerships/?cid=FSEPRD592734.

3 Malamuth, Eldad, and Michael Mulheisen. "*Galucomys sabrinus* Northern Flying Squirrel." *Animal Diversity Web*, University of Michigan Museum of Zoology, 2011, animaldiversity.org/accounts/Glaucomys_sabrinus.

4 Nowacki, Gregory, et al. "The Current Status of Red Spruce in the Eastern United States: Distribution, Population Trends, and Environmental Drivers." *Proceedings from the Conference on the Ecology and Management of High-Elevation Forests in the Central and Southern Appalachian Mountains*, U.S. Forest Service, United States Department of Agriculture, 2010, www.nrs.fs.fed.us/pubs/gtr/gtr-p-64papers/15-nowacki-p-64.pdf.

5 Pauley, Thomas K. "The Appalachian Inferno: Historical Causes for the Disjunct Distribution of *Plethodon nettingi* (Cheat Mountain Salamander)." *Northeastern Naturalist*, vol. 15, no. 4, 2008, pp. 595-606. *Academic Search Elite*, doi:10.1656/1092-6194-15.4.595.

6 "Pigmy Salamander, *Desmognathus wrighti*." *Tennessee Wildlife Resources Agency*, www.tn.gov/twra/wildlife/amphibians/salamanders/pygmy.html.

7 Rosenberg, Matthew B., and David J. Butcher. "Investigation of Acid Deposition Effects on Southern Appalachian Red Spruce (*Picea rubens*) by Determination of Calcium, Magnesium, and Aluminum in Foliage and Surrounding Soil Using ICP-OES." *Instrumentation Science and Technology*, vol. 38, no. 5, September/October 2010, pp. 341-358. *Academic Search Elite*, doi: 10.1080/10739149.2010.508968.

SPECIES AT RISK: VIRGINIA BIG-EARED BAT

Jazmon Pennington

"It is simply amazing how quickly attitudes improve when people finally understand bats as they really are—sophisticated, beautiful, even cute, quite aside from their crucial roles as primary predators of insects, **pollinators** of flowers, and dispersers of seeds."

— Merlin D. Tuttle, *The Secret Lives of Bats*

The Virginia big-eared bat (*Corynorhinus townsendii virginianus*), a subspecies of the Townsend's big-eared bat, inhabits eastern Kentucky, northwestern North Carolina, southwestern Virginia, and eastern West Virginia. In 1979, the U.S. Fish and Wildlife Service listed this bat as an endangered species. Only about 18,000 exist, with most of them living in West Virginia. Bats are the only mammal capable of sustained flight. The bat's four finger bones are long and connected by two layers of skin that form a wing. The wing runs the length of the bat's body and attaches to the hind leg. The two hind legs are connected by a membrane called a uropatagium. Despite being one of the largest cave-dwelling bats in the eastern U.S., the Virginia big-eared bat is less than four inches (10 cm) long and weighs less than half an ounce (14 g). Its membranous tail, within the uropatagium, is nearly two inches (5 cm) long. It has small eyes but relatively good vision. The Virginia big-eared bat lives up to its name with nearly one-inch (2.5 cm) long ears.

Virginia big-eared bats live in groups called colonies. "They prefer caves in karst regions (areas with limestone bedrock) and **hibernate** in tight clusters near the entrances of caves that are well ventilated."[1] The two largest colonies spend winter in a cave on private land in West Virginia and in Stillhouse Cave in the Daniel Boone National Forest of Kentucky.

Figure 47 Virginia big-eared bat

Like all mammals, bats are warm-blooded. This means their bodies, when healthy, are capable of maintaining a constant body temperature regardless of temperatures outside. And since bats are mammals, mothers give birth to fully formed young and produce milk to feed them. The babies, called pups, cling to their mothers as the females fly around.

"The female Virginia big-eared bats separate into maternity colonies during the summer months, typically giving birth to one pup during June. The young pup grows rapidly and can fly within three weeks after birth. The young bats are typically weaned, nearly full grown, and feeding on their own at about seven to eight weeks."[6] "When a female bat returns at midnight after hunting for food, the female bat can find her pup by its own smell and call."[5]

Like most bats, the Virginia big-eared bat is mainly nocturnal, providing humans with natural pest control by feeding on insects. Moths make up the bulk of this bat's diet, but mosquitoes—which are not only pesky but can be deadly by spreading diseases such as West Nile virus—flies, wasps, termites, ants, and spiders are also consumed. The Virginia big-eared bat also hunts many species of agricultural pests such as June beetles, click beetles, leafhoppers, planthoppers, spotted cucumber beetles, green stinkbugs, and corn earworm moths. Similar to other insectivorous bats, the Virginia big-eared bat typically consumes one-third to one-half of its body weight in insects each night. "Published estimates of the value of pest suppression from bats averages about $74 US per acre; therefore, the value of bats to the U.S agricultural industry is about $22.9 billion US per year."[4]

Figure 48 Colony of Virginia big-eared bats

Bats are skilled hunters with unique abilities. "The reason they can fly around and hunt insects so effectively is because of their ability to use **echolocation**, or high frequency sound **navigation**. Using this highly evolved sonar system, bats make calls using their mouths or noses and listen to the echoes. Using sound waves, they can literally form images of objects as small as a human hair in their brains, much as we do from light waves reflected off of what we see."[3] Like those of most bats, the echolocation signals made by Virginia big-eared bats are too high-pitched for human hearing to detect.

In the winter, colonies of Virginia big-eared bats enter a long-term state of torpor, an involuntary form of hibernation. Before winter, the bats must develop a layer of fat just beneath the skin that protects against heat loss and provides nutrients during their winter sleep. If a bat wakes up in winter, its body temperature will need to rise, and it must use valuable resources to stay warm. With no available prey to hunt, the bat will starve. Researchers are particularly interested in the Virginia big-eared bat because this species appears to be immune to the effects of white-nose syndrome (WNS), a fungal disease that causes bats to wake up from hibernation.

While some Virginia big-eared bats have been found with white-nose syndrome on their bodies, the fungus does not typically manifest into the disease that leads to the death of the bats. Other species of bats around the world have been far worse affected by the disease, and many bat populations have plummeted—some by as much as 90 percent. It is unknown for sure how white-nose syndrome spreads and infects bats, and there is no known cure for this disease.

What is White Nose Syndrome?

"White-nose syndrome (WNS) is a disease that affects hibernating bats and is caused by a fungus, *Pseudogymnoascus destructans*, or Pd for short. Sometimes Pd looks like a white fuzz on bats' faces, which is how the disease got its name. Pd grows in cold, dark, and damp places. It attacks the bare skin of bats while they're hibernating in a relatively inactive state. As it grows, Pd causes changes in bats that make them become active more than usual and burn up fat they need to survive the winter. Bats with white-nose syndrome may do strange things like fly outside in the daytime in the winter." [7]

While the Virginia big-eared bat has been mostly spared from the terrible disease that is wiping out other bat species, this bat continues to be endangered, mainly due to human interference. Habitat disturbance, vandalism, and economic growth all threaten the Virginia big-eared bat. "Human disturbance is one of the largest factors contributing to the demise of these bats. Disturbing bats during hibernation can cause them to lose fat reserves, drop their young and, after repeated disturbances, even [die]."[1]

Bats can experience severe population decline when their habitat is disturbed. They instinctively return to the same caves every winter, places called hibernacula, which makes them vulnerable to changes that occur in the caves. Bats also roost in trees during warm weather—a behavior that also leaves them vulnerable to influences from nature as well as humans. Human disturbances can fall under a number of different categories from logging, mining, and road development to pesticides and pollution. The popularity of cave exploration, or spelunking, has also impacted bats' lives.

When disturbed, Virginia big-eared bats may completely abandon their homes. Conservationists and scientists recognize the threats and have taken steps to address them. While most known hibernation and maternity sites have been gated and fenced to prevent entry, cave exploration is popular, and security breaches have occurred. Individuals and industry need to be aware of how their activities affect bats. Greer Lime Company operates an open pit limestone quarry near Hellhole, a 28-mile (45 km) long cave system in Pendleton County, West Virginia. The Hellhole site is used as a hibernaculum by almost half of the total population of Virginia big-eared bats, as well as five other bat species—including the similarly endangered Indiana bat. Greer Lime Company is responsible for keeping the public out of the cave, and a network of sensors has been set up to signal if Greer's quarrying operations are felt inside the cave. Should the quarrying disturb the Hellhole bat population, the consequences could be devastating.[3]

Figure 49 People must drop down through the pit entrance of Hellhole on a rope; the man in the orange hat at the bottom of the image is a caver.

Wind turbines, while important to the reduction of our reliance on fossil fuels, have proven to be a serious hazard for bats. The Virginia big-eared bat has not been spared from this threat. "Dead bats have been found in the vicinity of wind turbines in the Appalachians. Aside from the merits of wind power and its potential as a viable source of clean energy, the turbines themselves are killing the bats."[2] Another threat to bats is the expansion of housing developments. When excavating areas for the developments, it causes the bats to become scared and fly away from nearby caves; this may also disrupt their torpor if they are too stressed to return to their hibernaculum. It is not uncommon for caves to be intentionally collapsed or filled in during land development. When caves are destroyed while bats are away, the bats could be permanently displaced.

Still more threats to bats include agricultural pesticides. These toxic chemicals not only reduce the amount of prey that bats need to survive but are also deadly to bats that consume insects that come into contact with the pesticides. Pollution in its many forms is a threat to all wildlife, and bats are no exception. When people leave behind trash that bats may consume, the results could be fatal. "Though the reasons for the unexpected bat mortality in the Appalachians are unclear, it's going to take a concerted effort to come up with a solution. Bats deserve a little human foresight on their behalf."[2]

Some notable actions are being taken to help Virginia big-eared bats, but none have been successful enough to be considered solutions. One initiative occurred in 2005, when Virginia's governor at the time, Mark Warner, signed legislation making the Virginia big-eared bat the state bat. This made Virginia the second state to adopt a state bat (the first being Texas with its Mexican free-tailed bat). As a state bat, the Virginia big-eared bat is legally protected from persecution.

Due to concerns about the spread of white-nose syndrome among bat populations throughout the country, the U.S. Department of Agriculture requests that cavers and cave owners work mindfully to reduce cave traffic and to remember the primary rule of caving: Pack it in, pack it out! Never leave anything in a cave. It is also requested that in areas where endangered bat populations are higher, no non-research cave explorations take place. The University of Kentucky published a series of directives designed to assist the public in responding to the bat crisis:

- Do not enter protected caves.
- Reduce the use of pesticides that end up killing the butterflies and moths on which bats feed.
- If you own property with a cave, learn the proper method of installing a gate at the entrance to prevent disturbance while allowing the bats to move freely.
- Educate people about the benefits of the bats.[5]

Figure 50 Virginia big-eared bat

The Virginia big-eared bat is endangered and needs our help to return to sustainable numbers. There are many actions that can be taken to provide potential solutions to the bat crisis: putting up gates to keep people out of bat caves, reducing the use of chemical pesticides, and providing legal protection to more caves and land that are bat habitats. As individuals, we can take small actions to help bats, too. Simply educating others about the wonders of bats can have an impact on the public perception of bats. The more people know about the Virginia big-eared bat's plight, the more people will be willing to help these amazing creatures. Bats are a valuable member of the food web. Without them, nature would be thrown out of balance in ways that we may not even be able to consider today. It is vital that we recognize the need to protect and conserve the Virginia big-eared bat for future generations. We can do it!

---

1 Bonini, Christina. "The Virginia Big-eared Bat." *James River Association,* October 2015, thejamesriver.org/the-virginia-big-eared-bat.

2 Bro Staff. "Are Bats Dangerous? Don't They Carry Rabies?" *Blue Ridge Outdoors,* 2004, www.blueridgeoutdoors.com/magazine/october-2004/are-bats-dangerous-dont-they-carry-rabies-naomi-matzner-charleston-wva.

3 Eichelberger, C., et al. "(*Corynorhinus townsendii virginianus*) Virginia Big-eared Bat." *NatureServe*, 15 July 2020, explorer.natureserve.org/Taxon/ELEMENT_GLOBAL.2.100716/CORYNOHINUS_townsendii_virginianus.

4 Owen, Sheldon. "Bats & Zoonotic Diseases." *West Virginia Extension Service,* April 2020, extension.wvu.edu/natural-resources/wildlife/bats/bats-zoonotic-diseases.

5 "Virginia Big-eared Bat." *Office for Environmental Programs Outreach Services,* College of Agriculture, Food and Environment, University of Kentucky, oepos.ca.uky.edu/content/virginia-big-eared-bat.

6 "Virginia's State Bat: The Virginia Big-eared Bat *Corynorhinus townsendii virginianus.*" *Virginia Department of Conservation and Recreation,* www.dcr.virginia.gov/natural-heritage/karst-stbat. Virginia Department of Wildlife Resources.

7 "Virginia's Endangered Rafinesque's Big-eared Bat." *YouTube*, 16 July 2008, youtu.be/Rw5rfAfcLFE.

8 "What is White-Nose Syndrome?" *White-Nose Syndrome Response Team*, U.S. Fish and Wildlife Service, Department of the Interior, www.whitenosesyndrome.org/static-page/what-is-white-nose-syndrome.

EFFECTS OF LIGHT POLLUTION ON WILDLIFE

Lori K. Ray

"For most of the wild things on Earth, the future must depend upon the conscience of mankind."

— Archie Carr, pioneering conservationist

Light pollution is a major environmental issue in our urbanized, industrialized world today. Light pollution, or the excessive use of artificial light at night, can be so intrusive and so intense that it virtually overpowers nature, disrupting photosynthesis, wildlife migration and reproduction, and a variety of other natural behaviors of plants and animals.

Humans didn't always need light. "The Earth used to have a rhythm of light and dark solely based on the sun, moon, and stars."[6] People once relied on the stars in the night sky for navigation, but now, in some parts of the world, it is difficult to get a glimpse of those same stars. As humans expand their reach farther into natural places, artificial light becomes a part of the human impact on our world. In fact, most of the world's population lives under skies polluted by artificial light.[6]

Artificial light can be seriously harmful to wildlife. Most outdoor lighting is completely unavailing because instead of illuminating specific objects or places, its broad aim illuminates wide swaths of land and sky, eliminating the natural darkness of the night sky and with it the normally visible cosmos. Artificial light that glows directly toward the sky causes wildlife—including plants—to become confused. For example, many species rely on natural light to travel and hunt. Artificial light may send them in the wrong direction, reveal their presence to predators, or prevent them from successfully hunting prey by stealth.

Jo Garrett, a researcher at the University of Exeter, reported that many biodiversity areas that are severely affected by light pollution are highly populated.[8] Birds, for example, that live in these areas have high rates of collision with vehicles and buildings. Birds have an inner compass that helps them navigate. Their bodies contain a protein called cryptochrome that allows the birds to sense light for the purpose of navigation. Artificial light and light reflections disrupt birds' natural ability to navigate. Scientific research has revealed that when "a particle of light, or photon, hits a bird's cryptochrome, its energy can perturb molecules within the protein."[10] This explains why birds are sensitive to light and why they can become disastrously overwhelmed by the energy coming from artificial light.

Cities use an incredible amount of artificial light. David Willard, a retired field ornithologist, has argued that reduction of artificial light in large cities, such as Chicago, could significantly reduce bird fatalities.[9] Birds that specifically navigate at night, such as owls, nightjars, and nighthawks, for example, are not accustomed to sensing a lot of energy from light sources, and exposure to artificial light can cause them to become highly confused, leading to lethal collisions with buildings.

Many bird species rely on daylight cycles to navigate as well as to breed, nest, and hatch. Light pollution can cause birds to hatch early in a season—a time that may not coincide with availability or abundance of prey. For instance, if insect populations are not ready to sustain the bird populations that prey on them, then the insects could be wiped out before having the opportunity to reproduce and rebuild their numbers.[7] Light pollution that shifts the natural behavior of one species most certainly leads to circumstances that affect the behavior—and even survivability—of others.

Light pollution from oil rigs and coastal development affects **zooplankton**, tiny animals that feed on drifting algae in the ocean. Research has shown that artificial light causes zooplankton to stop eating algae. When this happens, the imbalance in nature leads to algae blooms. Some of these blooms are toxic, resulting in the death of fish and marine mammals.[7]

Insects are also crucial to the natural balance of the environment. Among their many roles in nature, they serve as soil aerators, pollinators, and prey for other wildlife. For example, fireflies are one of the natural beauties of the night, but sadly, artificial light is dimming their **bioluminescent** magic and endangering them. Fireflies mistake artificial light for other fireflies, which adversely affects their mating and breeding. Research has shown that fireflies flash their bioluminescent lights less frequently around artificial light. This may leave many individuals without a mate, thus reducing entire populations of fireflies. Like all insects, fireflies are important to their ecosystems. They do more than just light up and look fun at night. A research team at the University of Virginia reported that fireflies decompose matter and are also an important food source for a variety of other insects.[2]

Figure 51 The more than 2,000 firefly species in the world are not considered endangered; however, their numbers have been steadily decreasing, and most researchers agree that development and light pollution are to blame for the decline of firefly populations.

*Figure 52 The cecropia (*Hyalophora cecropia*) is North America's largest native moth.*

Artificial light can also make insects more vulnerable because there is no darkness in which they can hide. Such exposure can easily lead to overpredation. Because lights attract insects at night, swarms that gather around artificial lights become a veritable buffet for predators who take advantage of the easy meal night after night. One notable group of insects that regularly fall victim to light pollution is the family of giant silkworm moths (Saturniidae). Members of this family include the cecropia, luna, polyphemus, and promethia moths. With **wingspans** of up to six inches (15 cm), the North American Saturniids are the largest moths on the continent. These moths are beautiful, majestic creatures with colorful, intricate patterns and eyespots on their wings. Sadly, however, these unique moths are becoming less common for a number of reasons that include pollution, pesticides, **parasites**, and light pollution.[4]

We all know that frogs eat flies, and anteaters eat ants. But insects don't necessarily have to be eaten in order to influence the diet of other animals. Insects and other nighttime pollinators are driven away by artificial lighting. A research team from the Institute of Ecology and Evolution at the University of Bern found that pollinators visited flowers on meadows that were experimentally illuminated with street lamps about 60 percent less frequently than those that were on meadows without any light sources in the vicinity.[5] Further research is being conducted on this problem with the goal of trying to understand how much light pollution needs to be reduced in order to avoid shifts in the behaviors of many nighttime pollinators. When pollinators cannot do their job, trees and plants cannot adequately produce the fruit or flowers on which many other species depend. Every member of a food web—both plants and animals—is interdependent. Like dominoes, if one gets knocked down, all the rest end up falling in succession.

Plants are also affected by light pollution. Interestingly, plants were once used for navigation because they grow and move based on the location of the sun during the day. Tristan Gooley, author of the book *The Natural Navigator*, explains that "a flower that points to the sun is not adding much to the cause if the sun is clearly visible. A plant that can reveal where the sun has been over time, even at night or on an overcast day, is far more valuable for navigation."[3] Artificial lights can completely confuse plants, preventing them from revealing where the sun was at during the day. Since we no longer navigate using plants, this is not much of a problem; however, artificial lights also affect plant growth. Plants rely on sunlight for photosynthesis—getting nutrients from sunlight. Photosynthesis can be disrupted by light pollution, either by confusing the plant or damaging leaves so the plant or tree does not grow, flower, or bear fruit. Confused night-blooming flowers may not open under artificial lights, leaving wildlife that depend on their **nectar** without a food source. Plants are light sensitive, and when artificial light affects plant responses, other members of the plants' ecosystem could suffer.

*Figure 53 The Mexican long-tongued bat (*Choeronycteris mexicana*) is a nighttime pollinator that relies on cactus blooms so it can feed on the nectar.*

Figure 54 Artificial light on beaches can confuse sea turtles.

Case Study: Sea Turtles in Florida[1]

Florida is the only state in the continental U.S. where leatherback sea turtles regularly nest. The other sea turtle species that nest in the southeastern U.S. are hawksbill, green, loggerhead, and Kemp's ridley. Hatchlings emerge from their buried nests on sandy beaches at various times from June to October. Many people believe hatchlings emerge only during a full moon, but this is not true Hatchlings ready to emerge wait just beneath the sand surface until the day's heat subsides. When they emerge from the sand, instinct drives them toward the brightest light they can detect. On a natural beach, this would be the night sky reflecting on the surface of the ocean. But artificial light from homes and hotels on beaches are brighter than the night sky. The intense light lures the hatchlings away from the water and toward certain death. They exhaust themselves and perish on the dunes. They are attacked and eaten by animals such as raccoons, crabs, feral cats, and birds. They are struck by vehicles. And they never reach the sea.

Figure 55 Turtle-safe lamps light the Tarague Beach recreational area on Andersen Air Force Base, Guam. They are designed to emit light in lower wavelengths, which sea turtles are unable to see, giving hatchlings a better chance at survival by following the natural light from the night sky instead of bright artificial lights.

Figure 56 The beaches at Cape Canaveral Air Force Station, Florida, and Patrick Air Force Base located between Satellite Beach and Cocoa Beach are prime nesting habitat for loggerhead, green, and leatherback sea turtles. These communities are taking steps to address the issue of sea turtles and light pollution.

The problems associated with light pollution can be addressed by each and every one of us. People can close their blinds or drapes during the night to keep artificial light mostly inside their homes. Light coverings can be placed on outdoor light sources so that insects will not fly directly into the light and be killed. It is also helpful when people advocate for less artificial light use in their cities. A lot of people don't understand the amount of light pollution that artificial lighting can actually cause. Educating others about the issues of light pollution is also something we can all do.

Many organizations such as the Dark Sky Association, members of the scientific community, and even individuals are working to raise awareness of light pollution. Citizens are helping scientists collect data on how artificial light is affecting the brightness of the night sky in their areas. Many devices, apps, and maps have been useful in measuring the effects of light pollution. Phone apps such as the "Dark Sky Meter" and "Loss of Night" help people measure the brightness of the night sky in their areas. When citizens gather and share data, it helps scientists better understand areas that are affected by light pollution.

Light pollution can be addressed with human cooperation. While some individuals and communities are taking precautions to try to limit light pollution, not enough people are involved. Some people simply don't know enough about the harmful effects of light pollution, while others just don't seem to care. One of the easiest things that individuals can do to reduce light pollution is to simply turn off lights. Automated systems that turn off lights at specific times and motion-detecting lights that stay off until needed can help reduce light pollution. And some lights are designed to be low-glare, which also cuts down on light pollution.

Everything in the environment is interconnected. Our actions affect not only wildlife but also our own existence. Light pollution is something that many people don't think about, but it is a major problem in a lot of places. Let's all become more aware of light pollution and its impact on nature. By doing something as simple as turning off a light, we can take one simple step to protect the natural world from unnecessary disruption.

---

1 "Artificial Lighting and Sea Turtle Hatchling Behavior." *Florida Fish and Wildlife Conservation Commission.* myfwc.com/research/wildlife/sea-turtles/threats/artificial-lighting.

2 "Fireflies Need the Dark to Talk with Light." *International Dark-Sky Association*, 27 July 2017, www.darksky.org/fireflies-need-the-dark-to-talk-with-light/.

3 Gooley, Tristan. *The Natural Navigator.* E-book, Workman, 2012.

4 Johnson, Bill. "One Big Bug." *Horticulture,* vol. 115, no. 5, 2018, p. 61.

5 Knop, Eva, et al. "Artificial Light at Night as a New Threat to Pollination." *Nature*, 2017, doi: 10.1038/nature23288.

6 "Light Pollution." *International Dark-Sky Association*, www.darksky.org/light-pollution.

7 Macintyre, Ben. "We're Slowly Switching off the Sky at Night." *Comment | The Times*, The Times, 27 Dec. 2019, www.thetimes.co.uk/article/were-slowly-switching-off-the-sky-at-night-7qh93jnr3.

8 Marshall, Michael. "Light Is Leaking into Vital Habitats." *New Scientist*, vol. 241, no. 3219, 2019, p. 8. doi:10.1016/s0262-4079(19)30351-3.

9 Michigan, Jim Erikson. "Artificial Light Makes Birds Fly into Buildings at Night." *Futurity*, 5 Apr. 2019, www.futurity.org/birds-migration-artificial-light-collisions-2028542/.

10 Wu, Katherine J. "A Bird's Eye View of Quantum Entanglement." *PBS*, Public Broadcasting Service, 6 Feb. 2019, www.pbs.org/wgbh/nova/article/birds-quantum-entanglement/.

FROGS AROUND THE WORLD

Brogan Richards

"Come come! Come Out!
From bogs old frogs command the dark
and look . . . the stars!"

— Takarai Kikaku, Japanese haikai poet

Frogs are among Earth's many endangered species. Almost every frog species is affected by things that vary from place to place. For example, the purple frog of India is threatened by deforestation as well as harvesting and overconsumption. The lemur leaf frog, found in Colombia, Costa Rica, and Panama, suffers from habitat loss as well as a fungal disease called chytridiomycosis (known simply as chytrid). Ethiopia's Bale Mountain tree frog is under siege from habitat degradation caused by firewood collection and cattle grazing. Why should we care about little frogs? Because, as amphibians, frogs play a significant role in the balance of nature. Every species of frog fills a niche in its ecosystem. Frogs provide a valuable food source for larger animals, from snakes and birds to otters, raccoons, and foxes. As insectivores, frogs help keep insect populations in check. Frogs also play a role in the work of scientists, for a variety of important medical discoveries can be attributed to frogs. Almost all of Earth's frogs are in trouble. Their populations are declining—mostly due to human interference in their ecosystems. Scientists are racing to save dwindling frog populations on every continent they inhabit. Solutions exist for the protection and conservation of the purple frog, the lemur leaf frog, the Bale Mountain tree frog, and other species, but it will take serious action on the part of human beings to achieve lasting results.

What is an Amphibian?

An amphibian is a vertebrate animal with moist, scaleless skin of the class Amphibia, which comprises the frogs, toads, newts, salamanders, and caecilians. They are distinguished by having an aquatic larval stage, breathing through gills lined with frilly filaments that harvest oxygen from the water, followed by a lung-breathing adult stage. Amphibians are ectothermic animals. This means they cannot produce heat to warm themselves the way birds and mammals do. Their body temperature depends on the environment, so they must avoid extreme heat and cold.

Frogs inhabit every continent except Antarctica. Africa's goliath frog is the largest, at 35 inches (89 cm) long. The smallest four-legged vertebrates on the planet are two frog species found in Papua New Guinea. At 0.4 inches (10.1 mm) in length, it would take four of them to cover a penny. Let's look at some unique frog species around the world.

*Figure 57 Purple frog (*Nasikabatrachus sahyadrensis*)*

India's purple frog has small, golden colored eyes. Its dark purple skin covers a disproportionate and lumpy body. It is 2.5 to 3.5 inches (6–9 cm) long and typically weighs just under 6 ounces (170 g)—about the size and weight of a hockey puck. This frog is fossorial, meaning it is adapted to burrowing and living underground. Its burrows are three to six feet (0.9–1.8 m) deep. These frogs move around in loose soil to catch prey—primarily termites.

*Figure 58 Lemur leaf frog (*Agalychnis lemur*)*

The lemur leaf frog is a nocturnal rainforest dweller of Central America. Its skin is lime green during the day but darkens to brown at night. This amazing ability helps the frog hide from predators. Lemur leaf frogs range in size from about 1 to a little over 1.5 inches (2.5–3.8 cm). Females are larger than males. The lemur leaf frog population has dropped 80 percent just since 2010. Although some individuals seem to recover from chytrid, this species is not immune, and its population is severely at risk.

What is Chytrid?

Chytrid is a disease affecting at least 30 percent of the world's amphibians. It is caused by a fungus called *Batrachochytrium dendrobatidis* (or Bd) and is considered by some scientists to be the worst pathogen in the world. Similar to the way mold grows on bread, the fungus grows on an amphibian's skin, affecting the tissue and causing problems with the animal's body temperature regulation, respiration, and hydration. Chytrid may kill only some individuals within an amphibian population, or it may wipe out 100 percent of other populations. Chytrid fungus is believed to be spread through populations by direct contact between adults and larvae, or through exposure to infected water. No effective measure is known for control of the disease in wild populations.

*Figure 59 Booroolong frog (*Litoria booroolongensis*)*

The Booroolong frog is found only in the Eastern Highlands of New South Wales and southeastern Victoria in Australia. Its body varies in shades of brown; its belly is white. It grows to about two inches (5 cm) long. The introduction of non-native fish such as trout, which feed on Booroolong eggs and tadpoles, as well as the spread of chytrid fungus, have been cited as major causes of this frog's waning numbers. In addition, habitat loss, **sedimentation** of streams, and wetland and water pollution due to mining and agriculture, have all contributed to the decline of this frog's population and led to the species' classification as critically endangered.

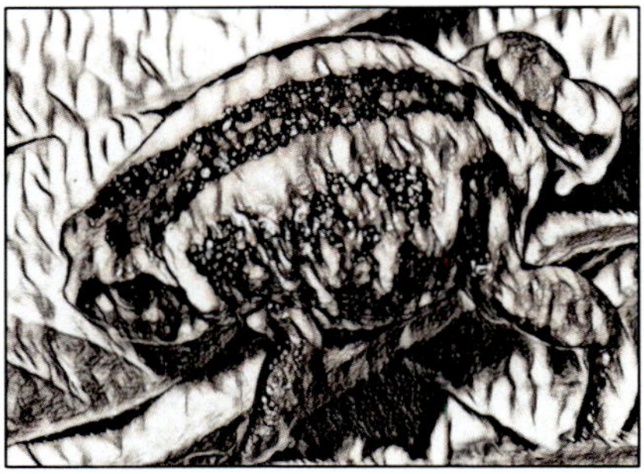

*Figure 60 Bale Mountains tree frog (*Balebreviceps hillmani*)*

The Bale Mountain tree frog is named for its habitat the Bale Mountains of the Ethiopian Highlands. These squat frogs are found at elevations of more than 9,800 feet (2,987 m). They are known to live in only one type of tree, the flowering tree heather. This frog has a round, strong body. It is dark brown to purple in color with golden stripes running down its back. Females average 1 inch (2.5 cm) long, while males average .08 inch (2 cm) long. They are so rare that very few images of them exist. Even though they live in the protected habitat of Bale Mountains National Park, their numbers are threatened by habitat loss and deforestation.

*Figure 61 Golden mantella (*Mantella aurantiaca*)*

Figure 62 Red form of golden mantella

Most of the roughly 300 frog species in Madagascar are found nowhere else on Earth. More than 20 percent of them are classified as threatened, and 9 species are classified as critically endangered by the IUCN Red List. The critically endangered golden mantella inhabits only a few isolated areas of the island's eastern forests. This poisonous frog is very small and has adapted two colors: golden and red. Not only is this frog threatened by habitat loss, but it is also captured for the exotic pet trade. Golden mantellas measure from 0.75 to 1.25 inch (2–3 cm) long, with females being larger than males. They spend most of the day hunting termites, ants, and fruit flies. Despite their brilliant color letting predators know that their skin is toxic, several species of snake prey on these frogs.

Frogs are vital links in the world's food chains because they function as both predators and prey. In the tadpole stage, they eat algae, helping regulate blooms and reducing the chances of algae contamination. Tadpoles also provide food for fish, birds, turtles, water snakes, and spiders. As adults, frogs help control insect populations while providing food for birds, fish, caimans, snakes, and small mammals such as various jungle cats and monkeys. Without frogs, the food chain could be affected forever.

In addition to being invaluable members of their ecosystems, frogs are also important to humans in the medical field. As humans become more resistant to **antibiotics**, other options are being considered. Scientists are using chemicals from frogs to develop new medicines. The *Smithsonian* website explains that samples taken "from their skin and from glands near their ears, called parotid glands, as well as from their bones and muscle tissues have been used as remedies for infections, bites, cancer, heart disorders, hemorrhages, allergies, inflammation, pain, and even AIDS."[2]

Human interference is certainly a major culprit in the decline of frog species around the world. As we push farther into wild areas to expand farms, cities, and resource exploitation, frogs find themselves pushed into ever shrinking habitats that often end up being contaminated with pesticides and other pollutants. Climate change is also to blame for the loss of frogs because it has been allowing the chytrid fungus to flourish. According to *Science News*, "warm periods enhance cloud formation over the tropics, which makes days cooler and nights warmer. Temperatures thus stay in the narrow range in which the fungus thrives, which could explain massive amphibian die-offs."[1]

Many herpetologists and other scientists are exploring ways to save frog species. These include the El Valle Amphibian Conservation Centre, commonly known as EVACC, located at El Nispero Zoo in central Panama. The center was set up and funded by the Houston Zoo. EVACC captures healthy frogs in the wild and then breeds them in **captivity** in order to keep critically endangered species from succumbing to chytrid fungus disease.

*Figure 63 Oregon spotted frog (*Rana pretiosa*)*

The Oregon spotted frog is Canada's most endangered amphibian. Habitat loss and predation by invasive species have led to a 90 percent drop in this frog's population. It is found only in small areas of British Columbia and Washington as well as along the Cascade Mountain range in Oregon, inhabiting lakes, wetlands, and slow-moving streams. In 1999, the Oregon Spotted Frog Recovery Team was created in order to protect this species from extinction. In recent years, the team has released more than 15,000 **captive-bred** tadpoles and juvenile frogs into the wild.

Unfortunately, many of the efforts to save frogs do not provide long-term solutions. Until a cure—or least a way to effectively manage—chytrid fungus is found, frogs will continue to suffer. If the frogs had a place to stay safe so they could repopulate, there might be hope for many species, but habitat loss, pollution, and chytrid contamination make this difficult if not impossible in most cases. Many things are causing frogs to be challenged in the natural world today. Frogs could go extinct if we do not take stronger action to better understand their needs and find effective ways to protect them and their habitats. While experts agree that more needs to be done to save not only frogs but all amphibians at risk, a first step that young people can take is simply to understand just how important frogs are to all life on the planet. Over 250 million years of evolution, there are today some 5,400 known species of frog, and they account for nearly 90 percent of all amphibians. In cultures all over the world, frogs are viewed as symbols of birth and rejuvenation, so it is ironic that frogs should be suffering so greatly under the threat of extinction. Let's all take a closer look at frogs so that we can help them survive in our changing world.

---

1 Brownlee, Christen. "Global Warming May Already be a Killer." *Science News*, vol. 169, no. 7, 8 Feb. 2006, www.sciencenews.org/article/global-warming-may-already-be-killer.

2 "Frog Toxins for Medicine." *Smithsonian*, 23 Jan. 2017, stri.si.edu/story/frog-toxins-medicine.

WHAT'S NEXT FOR THE POLAR BEARS?

Heather Sears

"The world cannot live without the Arctic; it affects every living thing on Earth and acts as a virtual thermostat."

— Philippe Cousteau, Jr., environmental advocate

When you think of the polar regions, do you think of ice? What do you think would happen if all that ice suddenly disappeared? How would everything that has depended on the ice for generations learn to live without it? Because of climate change, our polar regions are changing, and all the instinctual actions to harness the ice that polar animals used to survive for generations are now becoming useless. They must now either learn to adapt to their rapidly changing habitat or perish. Climate change and global warming are contributing to the breakdown of the polar regions, and by extension, have become the major threat to the future of polar bears.

Global warming is the accumulation of greenhouse gases that makes heat and essentially warms up the Earth. This is reducing the formation of sea ice and causing ice that does form to rapidly melt early in the season, thus reducing the polar bear's natural habitat to a fraction of what it was just a few decades ago. With the ice diminishing, there is little area for polar bears to hunt their natural prey: seals. Polar bears expend valuable resources trying to find ice—and prey—thus weakening until they finally succumb to starvation.

Historically, polar bears preferred to stay on the large sheets of ice that covered the region and enjoyed the shallow waters that the ice sheets rested upon. Now the ice is disappearing, and the water is deeper than the polar bears are comfortable with. The chance that polar bear populations can survive as they did for many generations is disappearing as fast as the ice.

Figure 64 Polar bears on the ice

Many things affect polar bears—from our exploitation of oil resources in the Arctic and its resulting habitat disruption to pollution to bear-human conflict and unsustainable hunting—but climate change is by far the greatest threat to polar bears. The main effect of climate change in the Arctic is the melting of ice. Although this may seem insignificant, the ice is a major element of polar bear life. These animals depend on the ice so that they can hunt for ringed and bearded seals, their main food source. Without ice, polar bears cannot hunt and must up their valuable **insulating** fat.

Polar bears are rather particular about the ice that they spend so much of their lives on. American zoologist Steve Armstrup pointed out that his team's research has shown that "polar bears preferred to live on the sea ice on shallow waters adjacent to the continental shelf."[1] However, due to climate change, the once shallow bodies of water that the ice formed on are now deep, warm pools. The vast ice sheets that stretched to the horizon in 1980 are now just small chunks drifting around the polar regions. This has limited the area that the polar bears can hunt, driving them to seek food far from the water. Polar bears now live in 19 sub-populations around the polar regions. Only two groups are known to be improving, while four are decreasing in health and numbers; five are considered stable for the time being, but the rest are a mystery as research has not been completed on them. While it certainly appears that some polar bears are adapting to changes in their environments in certain parts of the world, many other polar bears apparently cannot—they are suffering. Armstrup goes on to say, "By the middle of the century, we could lose two-thirds of the world's polar bears; and by the end of the century, we might lose them all."[1]

Figure 65 Polar bears struggle without adequate ice.

Many people do not believe climate change is a real event, or they are not concerned about it because they believe it is something that naturally occurs. However, to understand climate change, we need to look throughout Earth's past. There were numerous periods of extreme cold and warmth in the past that **fluctuated** quite a bit. Thinking back to the last ice age, you can see that the climate of the Earth has drastically changed at times before. However, as long as humans have inhabited the Earth, the climate has stayed relatively the same until recently. This recent change is alarming to many scientists because the change in the climate that we've been seeing just in recent decades is happening rapidly and extremely. And science has clearly shown that human actions have led to speeding up of climate change.

Will we be able to adapt to the rapidly changing climate in our own environments, or will we suffer as the polar bears are? This is the pressing question that is fueling scientists in their quest to help polar bears and by extension human populations. Steve Armstrup said in an interview that he and his team are "trying to save the world, one polar bear at a time."[1]

All the damage that we have been doing to the Earth—destroying forests and wetlands, poisoning air and water, driving wildlife to extinction, and overexploiting natural resources—is finally catching up on us. If we do not attempt to restore what we have destroyed, or at least remediate the damage that we are currently doing, then we are going to suffer just as polar bears are suffering. Earth's polar regions provide a means of measuring the health of our planet, and right now, the signs are not looking good for us. Our planet's health is rapidly declining.

Figure 66 What was once a vast sheet of sea ice is now an unstable, slushy patchwork.

Polar bears depend on ice, and the natural balance of the polar regions' animal population depends on the polar bears. Everything has a role, and everything depends on each other. When one piece of the ecosystem is removed or altered, then everything else in that ecosystem is affected. This holds true for the polar regions as well. With the ice disappearing, the polar bears have less places to live and hunt, the seal population soars while the polar bear population rapidly drops, and the balance of the region is negatively affected. Polar bears are needed to control the seal population because having too many seals in an area leads to overconsumption of fish, which then means the seals need to travel farther to find food. When seal mothers need to travel far, they lose weight and may be unable to nurse their young. Their pups are left vulnerable on the ice and may fall into the water and drown. Adults may be too thin to reproduce again. Without pack ice and fat seals, polar bears may need to swim farther out to sea or walk many miles inland to find food, thus expending energy reserves they can't afford.

Polar bear reproduction is also affected by the breaking up of ice. The polar bears are now broken up into small groups in scattered territories. Polar bears are now mating with brown bears. This could eventually lead to a loss of genetic uniqueness, and the polar bear as we know it today could cease to exist. As a species, polar bears are threatened or endangered in most of the regions where they exist. This is rather alarming because without them, the whole polar region's outlook appears rather bleak. If this fierce predator cannot survive, then the fate of the rest of the region is not as promising as researchers would like.

Figure 67 Ringed seal

Although too many people believe that climate change is not a major crisis, the condition of polar bears says otherwise. "Polar bears are a great example of what we're doing to the world in terms of global warming," says Steve Armstrup. "They're also a messenger about what's really important in addressing the threat."[1] The reason that scientists are focusing on the polar regions in their study of climate change is because these regions are showing more signs of climate change faster than the other regions. When the polar regions succumb to the effects of climate change, then the rest of the world will soon follow. Animals will starve, plants will die, and the human race will suffer an excruciatingly slow and terrible demise. Our polar wildlife is providing a warning for us to heal the Earth before it's too late.

Climate change is not the only threat to the polar bear population. Pollution is everywhere now, thanks to humans. From plastic in the ocean to trash thrown carelessly in campsites, we are doing a terrible disservice to the environment. Many animals—including polar bears—are hurt or killed regularly because of pollution-related issues. Residential and commercial development also challenges polar bears by taking away what little habitat they have left. With polar bears having to move inland to survive, and humans occupying or destroying the land that polar bears need, quite the conundrum has developed. Bear-human conflict is a serious problem. While only a handful of people have been killed by polar bears, many polar bears have been killed by humans. As climate change persists, polar bears will need to move farther inland and change their diets. If they can avoid humans, these bears could have a better chance at adapting to new habitats.

Figure 68 Polar bears may need to adapt to inland habitats.

In an attempt to save the polar bears, governments around the world have set aside land for polar bears, limited or banned hunting, and assisted researchers and conservation groups in studying and managing polar bear subpopulations. In addition, protected corridors assist polar bears as they move inland for survival. Although these efforts are helpful in the short term, they do not address the long-term problem of climate change, which has put polar bears in a precarious position in the first place.

When we look out of our windows, it is hard to imagine what irreversible damage we have done to the Earth. We can't see the depleting ozone or the disappearance of the large ice sheets in the polar regions, but it's still happening whether we see it or not. Everything we do on Earth that results in the mistreatment of our natural resources contributes to the deterioration of the health of our global environment—and us. Our greed and carelessness not only cause damage to the environment, but also to the living creatures that inhabit this world. There is no solution to erase what has already been done to the polar regions, but we can stop the situation from getting worse by changing how we treat the environment. Climate change is real, and it is serious. The damage that we are doing is drastically changing our Earth, and we need to address the situation with science, education, and serious effort before we reach the point of no return.

1 Tedx Talks. "Are Polar Bears Worth Saving? + | Steven Armstrup | Ted xUIdaho." *YouTube*, 1 September 2016, https://youtu.be/jX_R9ue4gyI.

THE CALIFORNIA CONDOR: A REMARKABLE RECOVERY

Paula Snyder

"We're seeing more chicks born in the wild than we ever have before, and that's just a step towards success for the condor and achieving a sustainable population."

— Tim Hauck, Condor Program Manager, Peregrine Fund

The California condor (*Gymnogyps californianus*) is a relic of prehistoric times. The great bird inhabited many areas of North America 40,000 years ago and is the only known bird species still in existence from the **Pleistocene**. Field biologists Noel and Helen Snyder noted, "As the largest flying bird of North America, and one of the most endangered, the California condor has been a source of tremendous interest and awe."[4] But modern human activity nearly wiped the California condor off the face of the Earth. It was only through the arduous efforts of committed scientists and conservationists that the California condor was able to recover from the brink of extinction.

Fossil records indicate the California condor's diet had evolved from consuming mega-sized vertebrate animals to large vertebrate animals, marine mammals, fish, and small birds. Fossils of prehistoric condors, known as Teratorns, had 14- to 20-foot (4–6 m) wingspans. Today's condors have a wingspan of up to 10 feet (3 m) and weigh up to 31 pounds (14 kg). Their large wings have distinctive white patches on the underside, and the wingtips resemble long fingers. California condors are the largest flying bird in North America. They primarily inhabit rocky cliffs in the central southern California deserts. Smaller populations are also found in Arizona, Utah, and Mexico.

Figure 69 California condor

With wings suited for soaring, condors can stay in flight for long periods of time in search of a meal. They are capable of soaring at heights of 15,000 feet (4,572 m). The California condor's existence in an ever-changing world can be attributed to its evolution from a predatory Teratorn to a much smaller **carrion** eater, for there's typically no shortage of food for scavengers. The California condor relies on its keen vision to locate food from above, dropping down to feast. Unlike those of a hunting bird of prey, the California condor's feet have adapted to walking, and their heads are featherless. This is an important characteristic, since they often dig their beaks into decaying flesh and would thus have trouble keeping a feathered face clean.

California condors remain with a single mate for life, but if one dies, the other will find a new mate. They nest on cliff faces, openings to caves, in rock crevices on steep slopes, or in burned-out tree cavities. The nests are not typical of most bird nests, though. California condors build nests that consist of gravel, stones, and bits of debris. A single egg is laid, and both parents participate in rearing the chick. Any number of things can go wrong in today's unstable environments, and they often do, which is why California condors have had such a difficult time maintaining their modern population.

No doubt it was its great size and majestic soaring that led early Native Americans to revere the California condor as a sacred bird, endowing it with the power of the sun. Native Americans used condor body parts to make valuable artifacts. When Europeans arrived in North America and pushed west, California condors were shot to obtain their hollow feather quills, which were used by prospectors as gold dust containers.[3] California condors were also killed for museum specimens and for sport trophies. Such activities were only the beginning of the human impact on these birds—an impact that nearly decimated their population.

As more settlers made their way across North America and built farms and ranches, they discovered that their livestock were being attacked by wolves, coyotes, mountain lions, and other predators. This led to the practice of poison baiting, by which a **carcass** would be filled with poison. While the poison killed the intended predators, it also killed scavengers such as California condors. Another method of pest control involved shooting predators with lead shot, which would then be consumed by the condors after the victims succumbed to their wounds. And in many cases, people simply mistook the condors for predatory birds and shot them on sight. Such practices continued for decades, and by the late 19th century, a sharp decline in the California condor population became evident.

By the early 20th century, formal protection for the California condor was introduced, with legislation making it illegal to kill the bird, collect the bird, and collect the bird's eggs; however, the inadvertent threat of poisoning the bird was not addressed, and the California condor population did not recover. Subsequently over the next several decades, advocacy volunteer organizations, conservationists, government agencies, naturalists, and wildlife biologists would study the California condor. These studies would include eating habits, nesting and fledgling activities, viability of the eggs, eggshell thinning issues, living environment, mortality rates, and a host of other aspects of the bird's life. While everyone agreed that the California condor's numbers were declining at an alarming rate, there didn't seem to be any real solutions in sight.

Figure 70 California condor wearing an identification tag as part of a conservation program.

Everyone had their own thoughts about how to study the bird, slow the mortality rate, increase the birth rate, develop formal sanctuaries and captive-breeding programs, offer food substation areas, and pass laws to require the use of nontoxic ammunition. The need to launch a successful

California condor conservation program was clear, for the California condor was on the cusp of extinction. Despite the recognition that California condors needed help, corporations continued to develop rows of wind turbines and power lines through the land that the California condor called home. Agricultural producers had been routinely using the deadly chemical dichlor-diphenyl-trichloroethane (DDT) as a pesticide since the 1940s and had dumped it in the Pacific Ocean for more than 20 years before the activity was halted in 1971. DDT polluted the soil and water and, when ingested, caused the shells of bird eggs to grow too thin. When the birds **incubated** their eggs, the shells would break. This led to plummeting bird populations—including those of the California condor. In addition, despite laws prohibiting the killing of California condors, a number of birds were shot by hunters who claimed they did not recognize the bird.

In 1967, the California condor was classified as critically endangered, but the population continued to decline. By 1980, preservation turned to panic. Conservationists began to capture the birds for their own protection. By 1987, the last wild California condor had been captured. The San Diego Wild Animal Park (later renamed the San Diego Zoo) and the Oregon Zoo stepped up to the challenge of trying to save the last remaining California condors. In April 1988, with only 27 California condors known to exist in the world, San Diego saw the first successful chick born in captivity.

*Figure 71 Condors lay only one egg per **clutch**, so every chick is precious.*

Figure 72 Puppets are used to rear chicks to prevent young condors from associating food with humans.

Numbers continued to fluctuate, and in 1992, the number of living California condors was down to 22. The first attempt to release captive California condors back into their natural habitat took place that year. However, by 1994, all of the released birds were recaptured because they were found to be at risk of failing to survive in the wild. For one thing, many birds were being electrocuted on power lines. The recaptured condors were then trained, with some of the training being **unorthodox**. One specific exercise was designed to keep the large birds from flying into power lines or wind turbines. Scientists in San Diego constructed false power lines in training pens that gave the birds a gentle electric shock in order to teach them that going near power lines would be unpleasant. Prior to this training, two-thirds of released birds were electrocuted by power lines, but after training, that number dropped to 18 percent. "Utility lines are not a significant problem anymore," said Bruce Rideout, a Doctor of Veterinary Medicine specializing in endangered species and lead scientist on the training program.[3]

Currently, Joe Burnett, a renowned wildlife biologist has dedicated most of his career to teach, preserve, and coordinate the California condor recovery program at the Oregon Zoo. Every year, Burnett and his team start setting up controlled condor nests, count birds, share data, and get excited with each success. Burnett says he cannot wait for the opportunity to see a bird he does not recognize.[2] The California condor population has now exceeded 450 living birds, with about one-third of those birds still in captivity.

In order to increase the number of potential chicks that can be produced and reared, a practice called "double clutching" is used. When a female condor lays a fertile egg, it is taken away by scientists, which induces the condor pair to lay another egg. The parents keep the second egg and raise the baby, but the first egg that was taken is artificially incubated, and when it hatches, the chick is raised by the scientists. Puppets that look like the chick's parents are used to feed the chick for a number of weeks. This helps the chicks associate with their own species. "It's kind of like pre-school for condors," says biologist Amy Utt of Loma Linda University, California, who works with the condors.[1]

But California condors that are released into the wild still face a variety of challenges, particularly toxic lead poisoning from ammunition used by people to shoot animals, which are often left to be cleaned up by the condors and other scavengers. Each year a number of California condors have to be recaptured and made to go through painful treatments due to lead exposure—and some birds do not live through the ordeal. In 2013, California addressed the problem of lead poisoning—the leading cause of death in released condors—by banning the use of lead ammunition. With the law having gone into full effect in 2019, conservationists are hopeful that their efforts to raise and release condors will not be in vain. Joe Burnett advises, "The single biggest thing people can do is, if they know or have family that hunt, or farmers, rancher family members, to get them to switch to non-lead ammunition. It's not only benefiting the condor, it benefits bald eagles, it benefits the whole ecosystem."[2]

The effort to save the California condor has been and will continue to be one of the most expensive alliances ever for a North American species. Education is a must if we are to continue to help the condor population grow and to protect birds released into the wild. Continued legislation to remove the use of harmful toxins in ammunition and limiting or eliminating chemical pesticides, as well as urban and industrial development planning to ease **encroachment** on the California condor's habitat are all measures that should be undertaken. We must stress the importance of conservation and preservation if this great bird is to survive and thrive for future generations.

---

1 "Kaplan, Matt. "Puppet Parents Raise Troubled Condors." Nature, 7 August 2007, www.nature.com/news/2C07/070806/full/070806-3.html.

2 "Saving Condors in Big Sur: A Q&A with Joe Burnett." *Oregon Zoo*, 4 June 2019, www.oregonzoo.org/news/2019/06/saving-condors-big-sur-qa-joe-burnett.

3 "Shock Therapy is Saving Endangered California Condor." *Discover Magazine*, p. 18.

4 Snyder, Noel F. R., and Helen Snyder. *The California Condor: A Saga of Natural History and Conservation*. Academic Press, 2000.

SEA TURTLES: NAVIGATING A CHANGING WORLD

Trey Waycaster

"The oceans are the planet's last great living wilderness, man's only remaining frontier on Earth, and perhaps his last chance to prove himself a rational species."

— John L. Culliney, marine biologist

Climate change has touched the existence of almost every living thing and nearly every aspect of our planet's environment. Roughly 84 percent of species on Earth are currently being studied in relation to the effects of climate change.[11] Included are loggerhead and green sea turtles, two important species of marine life that participate in preserving the health of their ocean habitats. They do so by maintaining productive coral reef ecosystems and seagrass beds. But sea level rise and climbing water temperatures—both direct results of climate change—are now threatening the survival of these amazing animals. Protecting and conserving our ocean resources is vital to not only the sea turtles' survival but also our own.

All seven sea turtle species are classified as either threatened or endangered. While the leatherback sea turtle is the sole member of the family Dermochelyidae, the other sea turtle species are in the order Cheloniidae, or those with hard shells. The loggerhead turtle is named for its relatively large head, which supports powerful jaws and enables it to dismantle hard-shelled prey, which includes whelks and conchs.[6] The green sea turtle is brown in color, not green. As the only herbivorous sea turtle species, it is named for its diet of seagrass, plankton, and algae. Green and loggerhead turtles may grow to be 450 pounds (204 kg) and 4 feet (1.2 m) long.

One of the greatest threats to all sea turtle species is increased ocean temperatures. The ocean absorbs vast quantities of excess heat from greenhouse gas emissions concentrated in the atmosphere (created mainly from fossil fuel consumption). "The Fifth Assessment Report published by the Intergovernmental Panel on Climate Change (IPCC) in 2013 revealed that the ocean had absorbed more than 93 percent of the excess heat from greenhouse gas emissions since the 1970s." We certainly haven't slowed down our emissions in the past few years, which means continued heat in the atmosphere is continuing to cause ocean temperatures to rise.[8] As the ocean waters warm, deoxygenation—a reduction in the amount of oxygen dissolved in the ocean—occurs. This leads to an imbalance in the chemical pH of the water—something that can have serious negative effects on marine organisms found at the bottom of the food chain as well as all those leading to the top. In addition, melting polar ice and the resulting thermal expansion of sea water has been causing sea levels to rise, which impacts animals that nest on coastlines—animals such as sea turtles.

Why is Water Temperature Important?
"Reptiles are ectothermic, meaning that their bodies depend on external sources of heat, and their body temperatures change with the environment. To warm themselves, sea turtles swim near the water's surface to absorb the heat of the sun, and when they need to cool off, they dive deep—as far as 2,300 feet (701 m)—to reach cooler water. Extreme temperatures are dangerous for sea turtles. When water temperatures fall below 58 °F (15 °C) in northern waters, sea turtles move toward warmer southern waters. On the other hand, when turtles become too warm, they seek out cooler climates. The most favorable body temperature for a sea turtle is about 75 °F (23.8 °C). If a sea turtle's body temperature falls below 40 °F (4.4 °C) for a day, it will die of hypothermia (being too cold). And if its temperature rises above 100 °F (37.7 °C), it will die of hyperthermia (being too hot)."[2]

The Sea Turtle Conservancy explains that "the direct impacts of sea level rise include losing beaches, ecologically productive wetlands, and barrier islands."[4] This means that sea turtle nesting areas are being affected. Sea turtles nest on land in summer. In the Northern Hemisphere, this is June through August, and in the Southern Hemisphere, this is December through February. Females come to shore at night and choose a spot that will be close to the sea but far enough from the water to remain dry. She uses her hind flippers to dig a hole about 17 inches (43 cm) deep.

Figure 73 Green sea turtle

Figure 74 Loggerhead sea turtle

Loggerhead and green sea turtles create up to four nests per season, leaving the ocean to create a new nest about every 14 days. Loggerheads lay 100 to 120 eggs each time, while green sea turtles lay up to 200 eggs per clutch. After all the eggs in a clutch are laid, the female turtle then uses her flippers to cover the hole and pat down the sand. Then she returns to the sea without ever knowing the fate of her offspring. But now it is getting very difficult for the sea turtles to find nesting habitats because of the rising sea levels. And many times, beaches continue to erode even while eggs are in nests, drowning the eggs or leaving them vulnerable to predators. These problems are happening all around the world, and we need to help find solutions before sea turtle losses are pushed past the point of recovery.

Rising ocean sea level and water temperatures aren't the only issues sea turtles face when reproducing. Rising atmospheric temperatures are a huge factor as well. Like that of most reptiles, the hatched sex ratio is highly determined by certain temperatures during the beginning of the incubation process. With sea turtles, in general, temperatures above 86 °F (30 °C) produce females and temperatures below 82 °F (27.8 °C) produce males. Since we are experiencing higher temperatures around the world as a result of climate change, more sea turtles are developing as female than male. At first, this may seem like nothing to be concerned about, since the warmer temperatures aren't leading to masses of unhatched eggs. However, consider this: as more females are being hatched, there could soon be fewer males available for reproduction. Since loggerheads do not mature for 15 to 30 years and green sea turtles for up to 50 years, such an imbalance in the sex ratio of their species could be disastrous. And as climate change persists, the environment may ultimately reach a point at which sea turtle eggs fail to hatch. Either eventuality would surely lead to the extinction of sea turtles.

Figure 75 Sea turtle eggs resemble the size and shape of ping pong balls.

Figure 76 Seagrasses are known as the "lungs of the sea." One square yard (~1 sq m) of seagrass can generate approximately 2.6 gallons (10 l) of oxygen every day through photosynthesis. [5]

Sea turtles are fundamental to the health of their ecosystems. Many corals have a relationship with photosynthetic algae. Protected by the corals, the algae produce oxygen and help the corals remove waste. But too much algae can smother the corals, killing them. Green sea turtles feed on the algae, preventing it from overgrowing. These turtles also keep seagrass ecosystems balanced. Seagrasses have roots, stems, and leaves that hold the plants in place. Some species produce flowers, fruits, and seeds as they grow on the ocean floor. Their leaves are used to spread nutrients and oxygenate the water—providing a healthy habitat for fish, crustaceans, and shellfish.[10]

Green sea turtles feed on the seagrass beds, which need to stay short in order to grow across the ocean floors and shores. The seagrass is actually what helps to keep the ocean clear and clean, which is extremely necessary for marine life to exist. Seagrass beds are where many young fish hatch. If the grass were allowed to grow too tall, then the fry would get tangled up and stuck in it. Maneuvering through the tall grass, foraging, and feeding would be far too exhausting, and wildlife would possibly die. In addition, seagrass that is too tall obstructs currents in shallow water, making water stagnant and toxic to the creatures that depend on this habitat. And tall seagrass creates too much shade for the seabed, so when the thick grass decomposes, it creates a layer of slime that suffocates the seabed and its inhabitants. Without green sea turtles to trim seagrass beds, the whole ecosystem suffers.

Figure 77 It is illegal to handle sea turtles in most places, but researchers may help babies escape from nests and carry them to the water in order to avoid being spotted by hungry seabirds.

Figure 78 Rescued and rehabilitated sea turtles may be fitted with temporary Global Positioning System (GPS) devices. Over a period of months or years, the devices send signals that can be picked up by satellites and used to track the turtles' movements.

Loggerheads also play an important role in the balance of marine life, particularly around coral reefs, where they tend to spend much of their time actively hunting crabs, mollusks, horseshoe crabs and other relatively large prey. When they feed on hard-shelled prey, they crush it and litter the area with the remnants of their meal. This not only feeds smaller creatures in the ecosystem, but it also "recycles essential nutrients and keeps ocean bed or floor sediments in complete balance."[12] Loggerhead sea turtles are themselves unique mini ecosystems in that many small plants and animals latch onto their shells and are carried around, benefiting from the movement by capturing food that drifts within reach. Research has revealed that "more than 100 species of plants and animals are believed to live on a single loggerhead sea turtle."[12]

Understanding that green and loggerhead sea turtles are so important in their ecosystems is the first step in recognizing that the loss of these animals would be devastating to the natural balance of the ocean. Rising sea temperatures are affecting seagrass and coral reefs, two ecosystems in which green and loggerhead turtles are crucially connected. "Seagrasses are the super ecosystems of our oceans, providing an incredible range of benefits to people around the world," says Dr. Maria Potouroglou, a seagrass scientist. Green sea turtles are a necessary component of these super ecosystems by not only relying on them for sustenance but also by managing the seagrass habitats. "Seagrass ecosystems are biologically rich and highly productive, providing valuable nursery habitats to more than 20 percent of the world's largest 25 fisheries. But an estimated 7 percent of seagrass habitat is being lost worldwide each year, and at least 22 of the world's 72 seagrass species are in decline."[3]

Coral reefs are incredibly valuable to the balance of the ocean as well. Roughly one-quarter of all marine species are found in, on, and around coral reefs. Loggerhead sea turtles, like green sea turtles, are both consumers and managers in coral reef ecosystems. But recent events induced by climate change have raised the alarm and threatened the future of coral reefs—and loggerhead sea turtles. In 2016 and 2017, heat stress affected 75 percent of the world's coral reefs, with 30 percent of reefs dying. Roughly 85 percent of the northern and far-northern Great Barrier Reef was struck particularly hard, with nearly one-third of the reef's shallow water corals being killed. In the Seychelles 99 percent of corals were bleached, and half died.[1] About 50 percent of the corals have been lost in the Caribbean region.[4] Scientists tell us that there may be a chance for some coral reefs to recover, but it could take decades to fully do so.

Climate change is certainly not the only threat to sea turtles around the world, but it is one of the most widespread, catastrophic, and deadly pressures that sea turtles face today. Human actions have exacerbated climate change, and human action is now urgently needed to work toward solutions.

Figure 79 Additional threats to sea turtles include commercial fishing and beach development.

In the face of enormous obstacles, actions are being taken to help sea turtles. Cooling chambers for the eggs bring down the temperatures to create a better chance for male sea turtles to be hatched, vegetation planted along dunes reduces erosion and thus improves turtle nesting sites, and marking and fencing of nests prevents beach goers from inadvertently disturbing them. Fencing is even placed over top of nests to keep predators such as raccoons, feral cats, and seabirds from digging up nests.

A problem that affects nesting and hatching is caused by artificial lighting on beaches. Switching to special lights that emit a light range that turtles can't see or just turning off lights along the beach can help prevent adult sea turtles from becoming confused when nesting and baby sea turtles from becoming confused when hatching and trying to make their way to the sea. And some of the simplest things that individuals can do to help sea turtles include picking up trash that has been left along the shorelines and in the water, flattening sand castles before leaving the beach, and filling in holes that people dig in the sand. These actions help sea turtles because if a female sea turtle comes ashore to nest and encounters any sort of obstacle—such as trash, a mound of sand, or a hole in the sand—she will turn around and return to the water without laying any eggs. Moreover, she may not return all season. It's important to keep the pathways open and clear for sea turtles to come ashore to nest. Also, if people encounter sea turtle hatchlings on the beach, the turtles should never be touched. But in order to give them the best chance of survival, people could follow them on their frantic race to the sea in order to provide them some protection from crabs, seabirds, and other predators.

Social media is a great way to share information about climate change and offer advice on what can be done to help sea turtles. Another powerful and easy solution is to simply tell a friend what you know about climate change and what we can all do to help sea turtles and other wildlife survive in this changing world. Every little bit of effort counts.

Something that is both fun and valuable is an idea launched by an organization called 4Ocean. They help out by selling bracelets that are made from recycled trash that was collected from the ocean. Their policy is "Buy a bracelet, clean the ocean." Visitors to their website (www.4ocean.com) can buy a bracelet to support ocean cleanup. 4Ocean has so far collected more than 9.8 million pounds of trash from the ocean.

Celebrities have also been stepping up to help support the work of scientists toward climate change solutions. Leonardo DiCaprio, Shawn Mendes, and Jane Fonda are just a few celebrities taking action. DiCaprio set up his own foundation in 1998, which is "dedicated to the long-term health and wellbeing of Earth's inhabitants."[9] His foundation has funded more than 200 projects in 50 countries around the world and awarded $100 million in grants. He even highlighted the need for everyone's involvement toward climate change solutions during a speech at the 2016 Academy Awards.

Acting legend Jane Fonda has been an advocate for climate change for a long time and has made headlines because of her Fire Drill Fridays, which she began upon being inspired by Swedish environmental activist Greta Thunberg's call to act like "our house is on fire" and guided by Canadian author, social activist, and filmmaker Naomi Klein and her Green New Deal advocacy. Jane Fonda has been arrested many times for leading rallies and calling for climate change research support and resources from the government. Fonda even pledged that her signature red coat would be the last piece of clothing she would ever buy—an action designed to help be more sustainable. In November 2019, at her fifth rally in Washington, DC, Fonda stated, "We cannot leave it to young people to fight this fight for their future by themselves."[7] Fire Drill Fridays continued to encourage people to gather online during the Covid-19 pandemic at firedrillfridays.com.

Exposure to the problem of climate change and its effects on the natural world is the best first step toward achieving solutions. Everyone needs to know what climate change is, why it's happening, and how it can be addressed. Education is key to achieving action. Let's all get involved in the fight to save the future of our planet!

Figure 80 Sea turtles are amazing ocean wanderers.

1 "Fast Facts: Coral Reefs." *Office for Coastal Management*, NOAA, 2020, coast.noaa.gov/states/fast-facts/coral-reefs.html.

2 Gish, Melissa. *Sea Turtles*. Creative Education, 2015.

3 GRID-Arendal. "Protection of Seagrasses is Key to Building Resilience to Climate Change and Disasters." *Phys.Org*, 8 June 2020, phys.org/news/2020-06-seagrasses-key-resilience-climate-disasters.html.

4 "Information about Sea Turtles: Threats from Climate Change." *Sea Turtle Conservancy*, 2020, conserveturtles.org/information-sea-turtles-threats-climate-change.

5 Jackson, Andrea. "Sea Turtles and the Environment." Sea Turtle Preservation Society, 20 July 2019, seaturtlespacecoast.org/sea-turtles-and-the-environment.

6 "Loggerhead Sea Turtle." *National Wildlife Federation*, www.nwf.org/Educational-Resources/Wildlife-Guide/Reptiles/Sea-Turtles/Loggerhead-Sea-Turtle.

7 Milka, Ruth. "Jane Fonda Leads Climate Protests at US Capitol for Fifth Week in a Row." Nation of Change, 11 Nov. 2019, www.nationofchange.org/2019/11/11/jane-fonda-leads-climate-protests-at-us-capitol-for-fifth-week-in-a-row.

8 "Ocean Warming." *Issues in Brief*, International Union for Conservation of Nature (IUCN), 2020, www.iucn.org/resources/issues-briefs/ocean-warming.

9 "Our Mission." Leonardo DiCaprio Foundation, 2020, www.leonardodicaprio.org.

10 Reynolds, Pamela L. "Seagrass and Seagrass Beds." *Smithsonian Ocean*, April 2018, ocean.si.edu/ocean-life/plants-algae/seagrass-and-seagrass-beds.

11 Witt, M. J., et al. "Predicting the Impacts of Climate Change on a Globally Distributed Species: The Case of the Loggerhead Turtle." *Journal of Experimental Biology*, vol. 213, no. 6, 26 February 2010, pp. 901-911. jeb.biologists.org/content/213/6/901#sec-3.

12 Yates, Brock. "Loggerhead Sea Turtle." *All Turtles*, 2020, www.allturtles.com/loggerhead-sea-turtle.

IMAGE CREDITS

Cover pxhere; Figure 1 Erin Thompson; Figure 2 Diane Maddox; Figure 3 Piqsels; Figure 4 NOAA; Figure 5 Mark Dixon; Figure 6 NOAA; Figure 7 NOAA; Figure 8 NOAA; Figure 9 Bernard Dupont; Figure 10 Nicole Kearney; Figure 11 Sardaka; Figure 12 Daniel Peláez Duque; Figure 13 Daniel Peláez Duque; Figure 14 U.S. National Park Service; Figure 15 ARC Centre of Excellence for Coral Reef Studies; Figure 16 Axis Travel; Figure 17 Kyle Taylor; Figure 18 Claudia Lombard; Figure 19 U.S. Fish and Wildlife Service Southeast Region; Figure 20 Brian Yurasits; Figure 21 TIO, electricteeth.co.uk; Figure 22 frontierofficial on Pixabay; Figure 23 Rabon David, U.S. Fish and Wildlife Service; Figure 24 Keyur Nandaniya; Figure 25 Adam Bignell; Figure 26 Maurits Bausenhart; Figure 27 John Clewley; Figure 28 N4RS Pixabay; Figure 29 David Mark; Figure 30 WikiImages; Figure 31 U.S. Department of Defense; Figure 32 Staff Sgt. Daniel J. Martinez, U.S. Air National Guard; Figure 33 Tom MacKenzie; Figure 34 Bill Greer, Florida Fish and Wildlife Service; Figure 35 John Fleck, FEMA; Figure 36 Steve Warren; Figure 37 ikinitip on Pixabay; Figure 38 Joe McKenna; Figure 39 pxhere; Figure 40 Joe McKenna; Figure 41 Pixabay; Figure 42 U.S. Fish and Wildlife Service; Figure 43 Knotnic; Figure 44 Brian Gratwicke; Figure 45 Billings Brett, U.S. Fish and Wildlife Service; Figure 46 Keith Kanoti, Maine Forest Service; Figure 47 Larisa Bishop-Boros; Figure 48 Craig Stihler, West Virginia Department of Natural Resources; Figure 49 Dave Bunnell; Figure 50 Jeff Hajenga, West Virginia Department of Natural Resources; Figure 51 yb_woodstock; Figure 52 Carmen Luna, U.S. Fish and Wildlife Service; Figure 53 U.S. Fish and Wildlife Service; Figure 54 ivansanchez3 on Pixabay; Figure 55 Staff Sgt. Melissa B. White, U.S. Air Force; Figure 56 James Rainier, U.S. Air Force; Figure 57 David V. Raju; Figure 58 frogmana; Figure 59 Tnarg 12345; Figure 60 Lois Kennedy; Figure 61 Cedricguppy; Figure 62 Frank Vassen; Figure 63 Mid-Columbia River Refuges; Figure 64 Brian McMahon; Figure 65 Jason Hillier; Figure 66 Wolfgang Hasselmann; Figure 67 NOAA Seal Survey; Figure 68 Dan Bolton; Figure 69 U.S. National Park Service; Figure 70 U.S. National Park Service; Figure 71 Joseph Brandt, U.S. Fish and Wildlife Service; Figure 72 Ron Garrison, U.S. Fish and Wildlife Service; Figure 73 Paul Asman and Jill Lenoble; Figure 74 Strobilomyces; Figure 75 P. Lindgren; Figure 76 Steve Hillebrand, U.S. Fish and Wildlife Service; Figure 77 South Carolina Department of Natural Resources; Figure 78 Andres Hernandez; Figure 79 Max Pixel; Figure 80 kormandallas

GLOSSARY

adapted – changed to improve its chances of survival in its environment
aggregations – clusters of things that have come or been brought together
amphipods – small, soft-bodied crustaceans with laterally compressed bodies
insidious – proceeding in a gradual, subtle way, but with harmful effects
antibiotics – medicines that kill or disable the growth of bacteria, or living organisms that cannot be seen except under a microscope
aquatic – living or growing in water
bacteria – microscopic living organisms, some of which cause illness or disease
biodiversity – the variety of life in the world or in a particular habitat or ecosystem
bioluminescent – having the ability to produce light from chemicals within the body
bioplastic – a type of plastic derived from biological substances rather than from petroleum and which breaks down through the natural process of decomposition
bushmeat – the meat of wild animals killed for food or for sale in tropical parts of the world such as Africa, Asia, and Australia
camouflage – the ability to hide, due to coloring or markings that blend in with a given environment
captive-bred – bred and raised in a place from which escape is not possible
captivity – living in a place from which escape is not possible
carapace – the hard, upper shell of turtles and tortoises
carcass – the dead body of an animal
carrion – the rotting flesh of an animal
chemical – relating to the interaction of substances that cause changes in matter
climate – weather conditions that remain mostly the same over a long period of time
climate change – the gradual increase in Earth's temperature that causes changes in the planet's atmosphere, environments, and long-term weather conditions
colonialism – the policy or practice of acquiring full or partial political control over another country, occupying it with settlers, and exploiting it economically
contamination – the action or state of making or being made impure by polluting or poisoning
clutch – one or a group of eggs produced and incubated at one time
commercial – used for business and to gain a profit rather than for personal reasons
crustaceans – animals with no backbone that have a shell covering a soft body
deforestation – the clearing away of trees from a forest
displaced – forced to leave one's home due to destruction or disaster
dung – waste matter eliminated from the body of an animal
echolocation – a system used by some animals to locate and identify objects by emitting high-pitched sounds that reflect off the object and return to the animal's ears or other sensory organs
ecological – having to do with the interdependence of organisms living together in an environment
ecosystem – a community of organisms that live together in an environment
edible – safe to consume as food
encroachment – movement into an area already occupied
endangered – seriously at risk of disappearing from the Earth forever

endemic – native to and confined to a certain geographical location
equator – the invisible line that stretches around the middle of the Earth
erosion – gradual wearing away of soil or rock by wind, water, or other natural agents
estuaries – the mouths of large rivers, where the tides (from oceans or seas) meet the streams
evolved – gradually developed into a new form
extinction – the act or process of becoming extinct; coming to an end or dying out
extirpated – made to cease to exist in a certain place
feces – waste matter eliminated from the body
feral – in a wild state after having been domesticated
fluctuated – rose and fell in an irregular number or amount
food web – a system in nature in which living things are dependent on each other for food
fossil – the remains or impression of a once-living organism preserved in rock
fossil fuels – fuels such as oil and natural gas formed from the remains of once-living organisms
fragmented – an organism's habitat having been broken up into scattered sections that may result in difficulty moving safely from one place to another
gene pool – the stock of different genes in an interbreeding population
genetics – relating to genes, the basic physical units of heredity
genocide – the deliberate killing of a large group of people, especially those of a particular ethnic group or nation
Global Positioning System (GPS) – a system of satellites, computers, and other electronic devices that work together to determine the location of objects or living things that carry a trackable device
greenhouse gases – gases in Earth's atmosphere that trap heat
habitats – the natural homes of plants and animals
herbivorous – subsisting on a diet of primarily grass, plants, and other vegetation
hibernate – to spend the winter in a sleeplike state in which body temperature drops and breathing, heart rate, and other body systems slow down
illicit – forbidden by law, rules, or custom
inbreeding – the breeding of individuals that are genetically related to one another
incubated – having kept an egg warm and protected until it is time for it to hatch
infrastructure – the basic physical and organizational structures and facilities (e.g. buildings, roads, power supplies) needed for the operation of a society
insulating – protecting from the loss of heat
invertebrates – animals that lack a backbone, including shellfish, insects, and worms
larval – of the newly hatched, often wormlike, form of many fish, crustaceans, and insects before they become adults
mammals – warm-blooded animals that have a backbone and hair or fur, give birth to live young, and produce milk to feed their young
manatees – aquatic mammals with a rounded tail flipper, living in shallow coastal waters and adjacent rivers of the tropical Atlantic
marsupial – a type of mammal, the female of which gives birth to undeveloped offspring that continue to develop outside the mother's body, in a pouch or fold of skin on the abdomen

meteorological – relating to the branch of science concerned with the processes and phenomena of the atmosphere, especially as a means of forecasting the weather
migrate – undertake a regular, seasonal journey from one place to another and then back again
militia – a military force that is raised from the civil population
mollusks – members of a group of spineless animals that includes snails, slugs, mussels, clams, and octopuses
navigation – determining a position and planning and following a route
nectar – sugary fluid produced by plants, especially within flowers to encourage pollination by insects and other animals
night vision – technology that changes light energy or detects heat patterns, making it possible to see objects in the dark
nocturnal – active at night
nomadic – relating to a group of people or animals with no fixed home who move, often according to the seasons, in search of food, water, or grazing land
nutrients – substances that give an animal energy and help it grow
organic – obtained from living matter (such as plants and animals) rather than chemicals or other artificial means
omnivorous – eating both plants and animals
parasites – animals or plants that live on or inside another living thing (called a host) while giving nothing back to the host; some parasites cause disease or even death
pathogens – bacteria, viruses, or other microorganisms that can cause disease
PCBs – chemical compounds once used abundantly in manufacturing and now illegal because of their toxic and long-term effects on living things and their offspring
pelt – the skin of an animal with the fur or wool still attached
photosynthesis –process in which green plants use sunlight to make their own food
Pleistocene – period in Earth's history that lasted from about 2,580,000 to 11,700 years ago; typically referred to as the Ice Age
poaching – hunting, trapping, or capturing protected species of wild animals, even though doing so is against the law
pollinators – animals or insects that transfer pollen from plant to plant, aiding in plant reproduction
predators – animals that hunt other animals for food
prey – animals that are hunted by other animals for food
primates – mammals with large brains and gripping hands; lemurs, monkeys, apes, and humans are primates
reptile – an animal with dry, scaly skin whose body temperature changes with its surroundings
tranquilizers – drugs used to induce relaxation or sleep
salamander – any member of the order Urodela, which are characterized by having long bodies, short legs, blunt snouts, and by living part of their lives in water and part on land
salinity – the amount of salt in water
satellite – a mechanical device launched into space; it may be designed to travel around Earth or toward other planets or the sun
sedimentation – process of soil, sand, and stone particles settling to the bottom of a body of water

species – a group of living beings with shared characteristics and the ability to reproduce with one another
stagnant – having no current or flow and often having an unpleasant smell as a consequence
sulfur – a yellow-colored non-metallic element found in various rocks and gases
sustainable – able to be renewed or kept functioning
symbiotic – a kind of relationship between two living things that is good for both
synthetic – a substance made by chemical or artificial means rather than by nature
territory – an area owned or claimed and defended from intruders
toxic – harmful or poisonous
trophy – object that is a symbol of victory or success
tundra – a flat, treeless Arctic region of Europe, Asia, and North America where the top layer of soil is permanently frozen
turbine – machine that produces energy when wind or water passes over its blades, which spin on a wheel or rotor
unorthodox – going against the usual ways of doing things
vapor – a substance diffused or suspended in the air, especially one normally liquid or solid
wingspan – the longest measurement across the wings of an object or animal, from wingtip to wingtip
zoologist – a person who studies animals and their lives
zooplankton – tiny sea creatures (some microscopic) and the eggs and larvae of larger animals